Lab Manual for Biomedical Engineering

Devices and Systems

Lab Manual for Biomedical Engineering

Devices and Systems

Third Edition

By Gary Drzewiecki

cognella®
SAN DIEGO

Bassim Hamadeh, CEO and Publisher

Carrie Montoya, Manager, Revisions and Author Care

Kaela Martin, Project Editor

Christian Berk, Production Editor

Jess Estrella, Senior Graphic Designer

Alexa Lucido, Licensing Manager

Natalie Piccotti, Director of Marketing

Kassie Graves, Vice President of Editorial

Jamie Giganti, Director of Academic Publishing

Cover image copyright © 2012 by Depositphotos / itsmejust.
       copyright © 2011 by Depositphotos / Малов Андрей.
       copyright © 2012 by Depositphotos / Dmitry Kalinovsky.
       copyright © 2012 by Depositphotos / Dario Sabljak.
       copyright © 2012 by Depositphotos / Viktor Levi.

Printed in the United States of America.

cognella | ACADEMIC PUBLISHING

3970 Sorrento Valley Blvd., Ste. 500, San Diego, CA 92121

Contents

Introduction to the Laboratory Course

OBJECTIVE

This set of lab exercises has been designed to parallel a one semester lecture course in biomedical systems and devices that follows a syllabus based on the book *Circuits, Signals and Systems for Bioengineers* by J. Semmlow. At Rutgers University, a lab-lecture course combination was designed to follow the lab exercises provided here. A syllabus can easily coordinate each book chapter with an equivalent lab exercise. This type of syllabus has been found to work well with Rutgers biomedical engineering students in that the lecture theory always precedes the lab exercise to help reinforce the lecture theory. In fact, it has been found that some students do not understand the lecture until after a lab exercise has been completed. This approach to teaching provides an additional value in that students can learn the first step in engineering design. That is, they understand that the theoretical concepts can be used to manipulate the lab outcomes.

COURSE COMPONENTS

In this lab course, there will be four components to every lab meeting:

1. A review of lecture material related to the lab exercise (**theory**). The following order of lab exercises will be found to correlate well with the Semmlow[1] text:

    01 Intro to Analog to Digital Conversion Equipment
    02 Wave Math
    03 Time Signals
    04 Time Frequency
    05 Noise Variability
    06 Fourier Transform
    07 Phasors
    08 Transfer Functions
    09 Thevenin
    10 Systems Modeling
    11 Amplifiers

2. A measurement-oriented exercise via digital data acquisition system (Biopac™ **experiment**). During this lab course, the student will learn how to configure the Biopac™ software; this allows for greater flexibility in experimental designs and also exposes the student to some key generalities of data acquisition systems. Hence, the student will be better able to adapt to other systems in the future, providing a broader knowledge base. Each experiment requires that the student generate a graphical output from both the digital records and the modeling software. Normally this data can be printed so that the student leaves the lab with "take-home" data that will be later analyzed for his/her lab reports. Most labs require the measurement of a model-parameter. The model-parameter is then used in the modeling component of the lab exercise and in the lab report write-up. Generally, this parameter will be identified from the classroom lecture theory.

3. **Modeling** via MATLAB™ Simulink™. In this portion of the lab, the student will learn to use Simulink to build a model that can be used to analyze the measurement portion of the experiment. Students will be able to work in the measurement environment and then return to the Simulink environment, a fundamental component of model building, and compare model results to the measurements performed. Students will further identify key parameters and variables of their model with the goal of obtaining the best representation of the experimental data. The philosophy behind this is to provide a chance for the student to complete a model design exercise on his/her own and then to examine how well it performs in comparison with real-world data. The student should return to revise his/her model design such that it is more representative of his/her data measurements. Each laboratory exercise should conclude with the student examining the relationship between his/her measurements and the computational theory representation of his/her data. Each exercise in this manual provides a set of sample data so that students can prepare for lab and also have an idea of what "good data" needs to look like. Finally, if necessary, students should make repairs to their model results to provide a better fit to data. Normally this only requires the adjustment of a few key model parameters. This repair process is useful because it also provides the student with an idea of how the theory represents real world data and may therefore be used to design the experimental outcomes.

4. **Lab reports.** The end of each lab experiment provides approximately 10 questions that the teams are asked to answer in the team lab reports. Normally, if students can successfully answer these questions, the lab report is considered complete. Instructors may choose to ask students to write more traditional reports that follow an outline of their choosing. Otherwise, instructors will find that the lab report questions provided here go towards further strengthening the lecture topics for each specific exercise. An additional set of questions is provided for individual lab reports.

Part I

Signals

LAB 1

Analog to Digital Conversion (ADC) Experiment

OBJECTIVE

In this experiment, we will explore the BIOPAC channel control parameters and data acquisition parameters that are available for adjustment to learn how to best achieve high-quality data recording. Also, you will measure the input noise present in your system and learn the basic calibration procedure. Then, the experiment will be modeled in MatLab™ by using the Simulink software.

KEY LEARNING SKILLS

- Calibration
- Linear systems
- Noise
- Data sampling
- Aliasing

EQUIPMENT LIST

- BIOPAC system connected to a PC and BIOPAC software
- Simulink software
- BNC-to-BNC straight-through adapter
- BNC to clip lead cable
- BNC T-Connector
- Calibrated DC voltage source (5 volts)
- Waveform generator

A. PROCEDURES

1. Calibration

Start by connecting the BIOPAC BNC cable connector to one of the BIOPAC system channel inputs. Use the BNC adaptor to connect the BIOPAC BNC cable end to the BNC to the clip lead cable. Connect the clip leads to the 5-volt output of the power supply. Turn on your supply. This provides a source of known constant 5 volts for calibration. You may need to attach some wire to the 5-volt terminals in order to attach the clip leads. You will next do an ADC calibration procedure. This is important to learn because you will need to do it almost every time you use the BIOPAC to measure voltage.

Start the Biopac Student Lab (BSL) Pro software by double clicking its icon. After the window opens, you should see a green light in the lower right. If not, ask for help; your system is not communicating with the computer. Then open the MP36 menu to select the channel to which you have connected the BNC cable to acquire, plot, and view the data. Select create/record new experiment, then select empty graph and click ok. Click on the MP36 and then the set-up channels tab. Put a check in the boxes labeled: acquire, plot and value. Select the type of sensory output connected to the BIOPAC under the preset menu; for this lab you would select BNC cable +10/–10 volts.

Then select the set-up button at the top right and choose a gain of 200x. This electronically sets the BIOPAC amplifier to multiply the input voltage on the channel by 200 before digitizing it. If you used the BIOPAC before, this set-up procedure was done for you. We have chosen a 200 times gain this time because the input voltage will be on the order of volts and does not need any extra amplification.

Verify that you have a known 5-volt DC source and connect it to your channel input.

Click the **scaling** button at the bottom of the window or the wrench icon at the bottom right of the graph screen. Go to MP36 menu to set-up channels, parameter, and scaling. This places 0 volts on the channel input. Type in 0 and 5 volts into the window for **cal values**. Click the CAL button next to the 5 volts.

Next, remove the cable from the 5-volt source and short the input clips together. This gives you 0 volts input. Now with the shorted input, click the CAL button next to the 0 value in the scale window.

Verify your calibration by opening the show input window on the MP36 menu.

You may use the show input menu to see the current voltage at the cable input.

Reconnect either 5 or 0 volts. Make sure that the show input is not on hold, otherwise it will keep showing the last input value. If this reading is correct, your calibration procedure is done. Note that we did NOT do a time calibration. This is because the MP36 unit contains a very accurate internal clock, so we need not worry about the time calibration. Now, record a voltage by shorting the cable again for 0 volts input. Then open the MP36 acquisition menu. Choose record once to memory. Enter 30 seconds duration and 100 Hz sample rate. Click start at the lower right. You should see a 0 volt constant line. Of course, the input is not truly 0 because we know that noise is present in the system. Next, measure the amount of noise in your system.

2. Measure Baseline Noise

After completing the calibration procedure, make sure that your cable input clips are still shorted together to provide a 0-input voltage to the MP36. Start another recording to verify that the MP36 indicates a constant 0 voltage. While displaying the time plot, click the vertical and horizontal resize buttons in the upper left. It may be visible that the voltage is actually fluctuating. This is the electronic noise present in your system. Without doing any other changes, go back to the channel input window and alter the channel gain from 200 to 5000. Record the time signal again for the cable shorted, for 0 volts input. Record and print your findings. You can autoscale the vertical axis if needed to see the noise better. Refer to Figure 2 at the end of this experiment for a sample Noise recording. If you see a constant line, you will need to go back to increase your acquisition rate to 1000 Hz in the MP36 acquisition menu. Record your measurements in the table below using the BSL *PRO* measurement cursors to do this.

SHORTED VOLTAGE INPUT NOISE MEASUREMENT USING MEASUREMENT CURSORS

Channel	P-P	Max	Min	Mean	Stddev
CH1					

The above data represent baseline minimum voltage that your instrument can measure. You cannot measure lower than this. It is the **error** that is always present in your data for future experiments.

3. Changing Acquisition Rate (effect of sampling rate)

You will now record a 2 Hz sine wave signal at a **high** and **low** acquisition rate to observe the effect on the sampled data. A perfect recording should yield a sine wave.

Remove the clip leads cable and connect the BNC input cable directly to the output of the function generator. Set the function generator to be in free run mode with no output attenuation, set the frequency multiplier to X1, and dial in 2 Hz. Push the waveform selector for a sine wave function. Return the channel gain to X200. Go to the acquisition window and set the acquisition rate to be 500 Hz (high sample rate). Record for a duration of 15 seconds. At the plot window click the start button to do a data recording. You should see a sine wave being plotted. If not, try to work until you get that result by verifying your settings. After you are successful, go back to the acquisition window and alter the acquisition rate to be 1 Hz (low sample rate). Start another recording. Make notes and print what you find for your decreased acquisition frequency. At this point you should have recorded two 2 Hz sine waves, one at 1 Hz another at 500 Hz acquisition rate. Recall that **sample time = 1/(acquisition rate)**. Record the sine wave measurements below for the 500 Hz sample rate and gain = X200 using the measurement cursors. These settings should have provided you with the most ideal looking sine wave in comparison with Figure 1. If not, work until you can achieve the best possible recording.

SINE WAVEFORM MEASUREMENT—GAIN = 200 AND F ACQ = 500 HZ

Channel	P-P	Max	Min	Mean	Stddev
Ch1					

After you have obtained the settings for the best continuous sine wave, go back to increase channel gain to X5000 and repeat a recording and record the measurements in the table below. Do not change any of your generator settings. You should notice that your sine record is no longer ideal and that the high gain situation has flattened the top or bottom of the sine waves. Print this for your records. This flattening result is defined as **clipping**. Fill in the table using the measurement cursers as you did above.

SINE WAVEFORM MEASUREMENT—GAIN = 5000 AND F ACQ = 500 HZ

Channel	P-P	Max	Min	Mean	Stddev
Ch1					

You will notice that the data has changed from the 200-gain case due to clipping. Thus, clipping is not a desired condition since it introduces large errors into the measurements. In future experiments you need to identify the occurrence of clipping and correct for it by reducing the gain factor.

B. SIMULINK MODELING PROCEDURES

Basics

Find the Matlab icon on your desktop and double click it to open the Matlab window. It is easiest to start Simulink by clicking the Simulink button at the top of the window. (File New—Model starts a new Simulink workspace in a new window.) Note, this is not the same as the command window. It is a new window specifically for building Simulink models.

Libraries

On the left of your Simulink window you should find the directories to the various Simulink function blocks.

A. Model Acquisition Rate Effect

Click to open the continuous signals library. Then click on sources, find the sine wave, and drag it to the untitled Simulink workspace. This should open a sine wave block in the Simulink window. Also locate the scope block in the library and drag it into your Simulink window. You can now use the mouse and cursor to draw a line connecting the output of sine wave block to the scope input. You will now enter the parameters that describe the sine wave that was recorded in your experiment. Double click the generator block to open its parameter window. You should then set it to output a 2 Hz frequency. Recall that period is inverse frequency and 2 pi f = angular frequency for Simulink. It is also useful to know that (peak-to-peak)/2 = amplitude. Run the model and observe the scope results. Now duplicate your BSL *PRO* recordings of the sine wave by changing the generator frequency while keeping step size constant. The step size corresponds to the reciprocal of the acquisition frequency. So, model the two experiments above **high** acquisition and **low** acquisition. Print these graphs. In this experiment we will not examine the effect of channel gain since the Simulink is a numerical nearly ideal computational system and does not suffer from the non-ideal aspects that a real-world amplifier system does. Your two model graphs should compare well to the BIOPAC data. If not, make repairs as needed. This close correspondence between model and experiment is an important aspect of all future experiments. It will assist you in understanding the course theoretical concepts and also, since all experiments will be linear or very nearly linear, the model should predict the experiment to a high degree of accuracy. The student should understand that this is the main element of engineering design!

Group Lab Report Questions

1. Provide copies of all plots: a) baseline noise record, b) high-acquisition rate sine BIOPAC data and sine model, c) low-acquisition rate BIOPAC sine data and model.
2. What was the channel baseline fluctuation voltage for the high gain recording? Describe the appearance of the noise. What kind of noise is it?
3. What was the mean (peak-to-peak) sine wave voltage observed with high gain conditions?
4. What is the theoretical v mean of a sine? Compare to your measurement. Did you find the correct value? If not, explain.
5. What voltage should you theoretically measure when the input leads are shorted together? Explain why.

6. What was the difference between your expected voltage and the voltage actually recorded when the input was shorted? Explain how this will impact your ability to measure biosignals. Provide physiological examples.
7. Discuss your observations for the sine wave recording while you varied the frequency. Make note of your results when you are ≥ or ≤ the acquisition rate.
8. What is your conclusion for best recording parameters of the sine wave?
9. What conditions in #7 yield nonlinear results?
10. Calculate the signal/noise for your sine wave and noise data. Express your result in dB and linear values.

Individual Lab Report Questions

1. The BSL *PRO* recording system has an ADC with a range of +/−10 volts. This requires that the input signals are shifted and scaled to fit within this range. Design and draw a sketch of a system that can perform this operation on a signal before it inputs to the ADC (you may use those system blocks that were introduced in Chapter 1). Note, this same system will provide voltage calibration of the 0 to 5-volt calibration voltage before it inputs to the ADC.
2. Go to the BIOPAC website to find the BIOPAC noise level specifications as measured by the manufacturer here: https://www.biopac.com/wp-content/uploads/MP_Hard-ware_Guide.pdf.
    a. Compare your measured noise levels to the BIOPAC specifications.
    b. Discuss the types of noise that you observed from your noise recordings in comparison to those presented in the textbook.

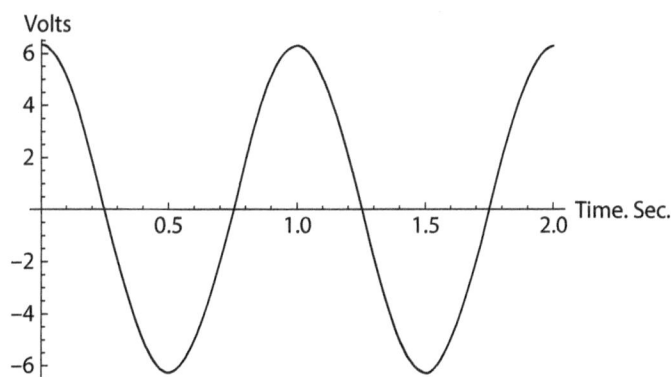

FIGURE 1.1   Ideal sine wave.

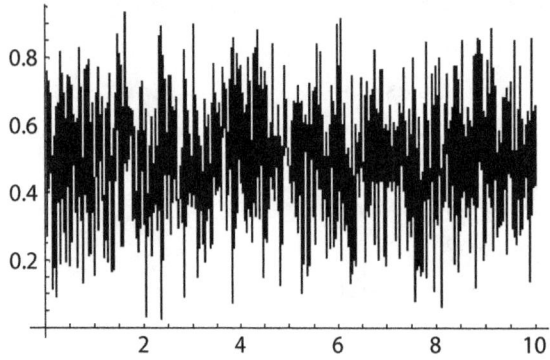

FIGURE 1.2 Sample baseline noise data.

Sample Lab Data

LAB 2

Waveform Math

OBJECTIVE
In this lab, we will record three different waveforms and apply math operations to them in Biopac and Simulink.

KEY LEARNING SKILLS
- Sine waves
- Linear math operations
- Sine measures

INTRODUCTION
Our prior lab focused on the recording of a simple sine wave voltage source. A sine wave was a useful signal in this lab because we are very certain of the shape of this function. Not only is it a useful function to apply in biomedical systems, but it also ensured the student that the BSL *PRO* data recording system was operating in a linear manner and not altering the data in any way. The student is now confident of the proper settings of his/her recording system such that his/her data are true, and the recording system is linear in operation.

Knowing that BSL *PRO* is a linear recording system, it is now useful to examine the effect of mathematical operations that can be applied to biological data. In this lab, we will focus on the effect of the calculus operations of differentiation and integration on a sinusoidal voltage. It is very common to apply these operations in biomedical systems theory. Working with a simple sine wave, a sine voltage is given as, shown in Figure 2.1.

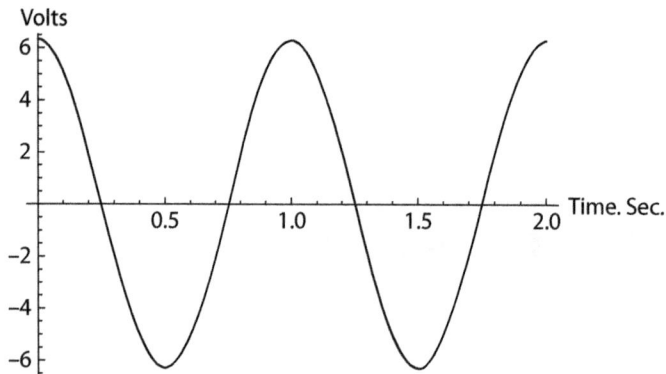

FIGURE 2.1 Sample Sine Wave of Amplitude A=1 and ω = 1.

A square pulse waveform oscillates periodically between constant positive and negative voltage for the duration of half the total period. Thus, if the integral is applied to a constant voltage, we find that the result is a line with constant slope in time where the positive voltage results in a positive slope and negative voltage results in a negative slope. Hence, we see the triangle wave as a signal alternates between positive and negative sloping lines in time or simply the integral of the square wave signal. This is exactly what your waveform generator is doing in its internal circuitry. In this lab experiment the math operations will be performed on your data recorded waveforms to prove this concept to you. Be aware that other more complex waveforms may be designed in this way and that you are not limited to the waveforms supplied by the Tektronics generator for your future experiments.

Next, we can perform the calculus derivative of the voltage v(t) as v'(t)=. The graph of this derivative is provided in Figure 2.2. It can be seen that the derivative voltage is earlier than the original or it is defined to be a phase lead. In this lab, BSL *PRO* will be used to find the derivative of a sine wave voltage as well. Be clear that in the lab we are working with a real-world voltage that is changing in time as a sine function. BSL *PRO* software will use an internal algorithm to calculate the derivative of the sine wave for you. The other voltages that can be generated by the wave generator will be analyzed in a similar way. You can see that real-world voltages can be manipulated mathematically by means of software calculations just as you are performing the mathematical analysis in your systems theory course. The laboratory data can be analyzed in an equivalent way.

EQUIPMENT LIST
- BIOPAC MP36 system connected to a PC and BSL *PRO* software
- Simulink software
- Bipac BNC cable
- BNC to BNC straight through adapter
- BNC to clip lead cable
- BNC T-Connector
- Waveform generator

Voltsderivative

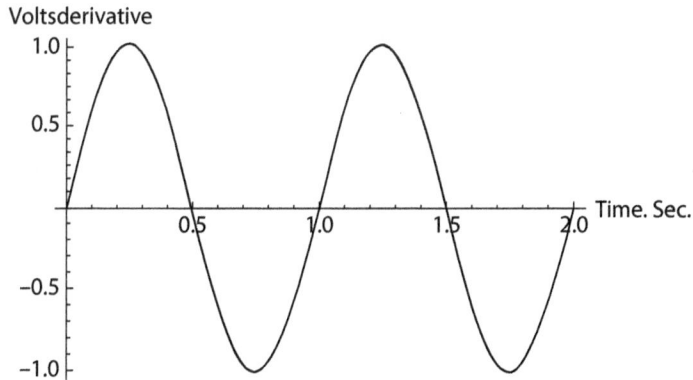

FIGURE 2.2 Derivative of the Voltage v(t) in Figure 2.1.

BIOPAC EXPERIMENT

Procedure

Calibrate channel 1 of the MP36 input voltage. Remove your 5-volt calibration source. Then, connect the BIOPAC BNC cable to just the waveform generator output. Acquire a 2 Hz sine waveform from the function generator. Verify that the wave is correct and that you have chosen the best acquisition rate.

1. Measuring Derivative and Phase

Record another 2 Hz sine wave. Then using the edit menu, duplicate the recorded waves. You will now have two identical tracings of sine wave. Select the duplicate sine wave and use the transforms menu to perform its derivative. You should now have the original sine wave and its derivative tiled in the same window. Use the measurements tool and cursor to find the peak-to-peak voltage and frequency of each sine wave. Print and record these data. You should see that one sine wave is shifted compared to the other. Find the time shift by using the delta T measurement with the cursor tool. Use the cursor measurement to find the frequency of the sine waves. Repeat the same experiment for the sine integral. The duplicate waveform step is used for calculations so that it can be compared to the original wave on the same window. Sample waves are at the end of this lab. Repeat your sine experiment for a square wave and triangle wave at 2 Hz. Use the waveform selection button on the generator to switch to these waveforms. Do not do phase measurements again. Note that the square wave derivative may be clipped since it is so large. Sample waves are provided at the end of this experiment.

2. Measuring Frequency

Next, change the acquisition rate to a lower level of your choice. Repeat the recording of a 2 Hz sine wave without changing the waveform generator. Then, measure the frequency of the sine wave using the measurement cursor. Does it match your previous value? If not, record the two different frequencies. Lastly, connect the generator trigger output to the frequency counter input with a BNC-to-BNC cable on your power supply. The set-up is provided at the end of this experiment section.

You should now have recorded four values of frequency for the 2 Hz sine wave:

1. Frequency counter value _____
2. Frequency from the wave generator _____
3. Frequency from measurement cursor (high acquisition rate) _____
4. Frequency from measurement cursor (low acquisition rate) _____

Read the frequency value from the counter.

SIMULINK

Use the Simulink program to model your waveform math experiment for the sine and pulse waveforms that you studied in the first part of this lab. You will need to create a new window for every calculation since Simulink does not allow the original and calculation in the same window. So, you will need to do integral and derivative on the sine and pulse separately.

Group Lab Report Questions

1. What was the peak-to-peak d/dt of the square wave and why is it limited? Is this the impulse function d(t)? (Refer to your textbook.)
2. Prepare a written summary of your experiment.
3. Provide all graphs—derivative sine, integral sine, derivative square, integral square—and same for Simulink.
4. Write the calculus equations for the derivative and integral of a sine wave.
5. Were your data accurate in comparison to the calculus values? Compare some data points to the theoretical values. Provide % errors.
6. Show that a triangle wave can be obtained from a square.
7. Discuss the differences between the Simulink results and your BSL *PRO* recordings.
8. How could you convert the pulse wave to a square wave?
9. Did you find that the frequency measurements match the generator frequency? Provide the percent error. How does the acquisition rate change the frequency measurement?
10. Find the accuracy of frequency in terms of the sample rate.
11. Calculate the phase shift between the original sine wave and its derivative and its integral from your data.
12. Does your measured phase match the theory? Provide both the phase angles and % errors.

Individual Lab Report Questions

1. Explain how you would use linear waveform math operations to convert a square wave into a triangle wave.
2. When you integrate a triangle wave you get a waveform that looks like a sine wave.
   a. Show that this sine wave is only an approximation. What is the true waveform? Sketch it.
   b. Explain how to double a sine wave amplitude by using differentiation. That is design a voltage doubler system.
3. Write the integral of the signal S(t) above. Sketch your result graphically.
4. Name three terms that describe the shift of the sine wave in this experiment.

Sample Lab Data

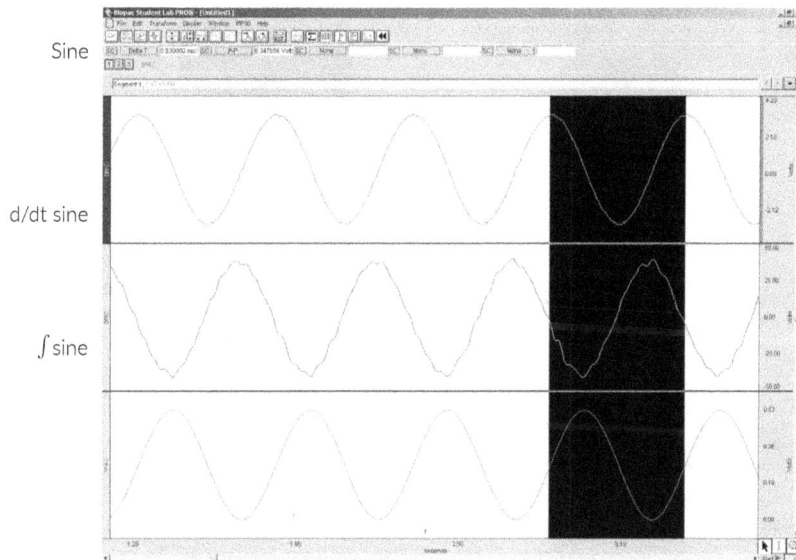

Sine Wave

Gain	X200
Acquisition rate	200 (Hz)
Period	0.500 (s)
Frequency	2.0 (Hz)
Vpp sine	6.348 (v)
Vpp d/dt sine	80.488 (v)
Vpp ∫ sine	0.532 (v)

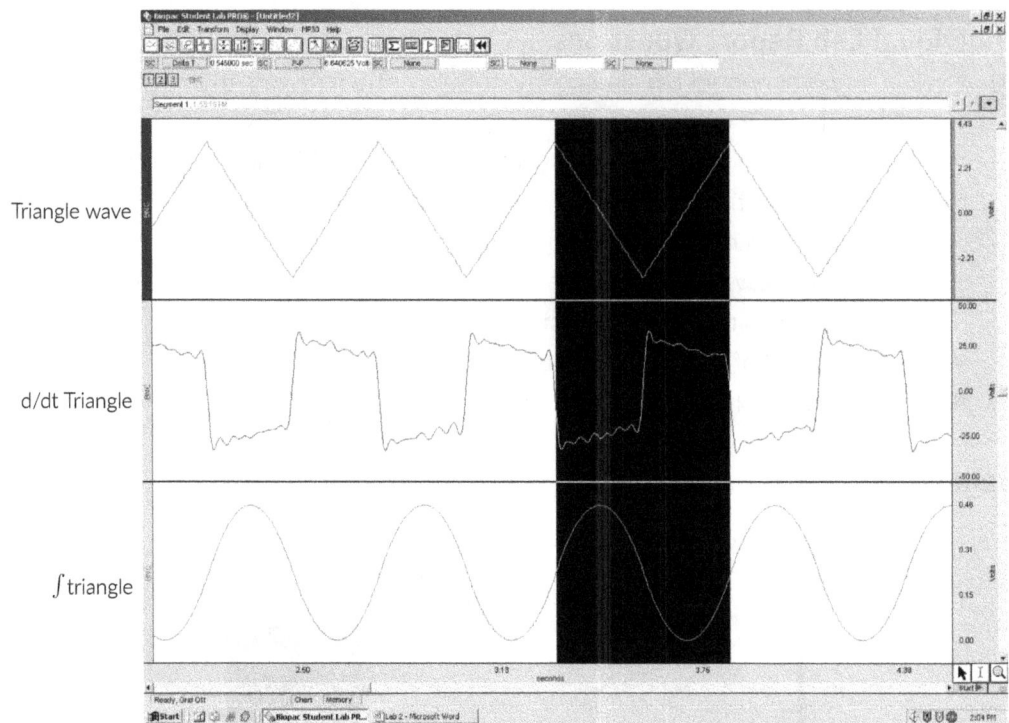

Triangle Wave

Gain	X200
Acquisition rate	200 (Hz)
Period	0.500 (s)
Frequency	2.0 (Hz)
Vpp triangle	6.641 (v)
Vpp d/dt triangle	65.824 (v)
Vpp ∫ triangle	0.458 (v)

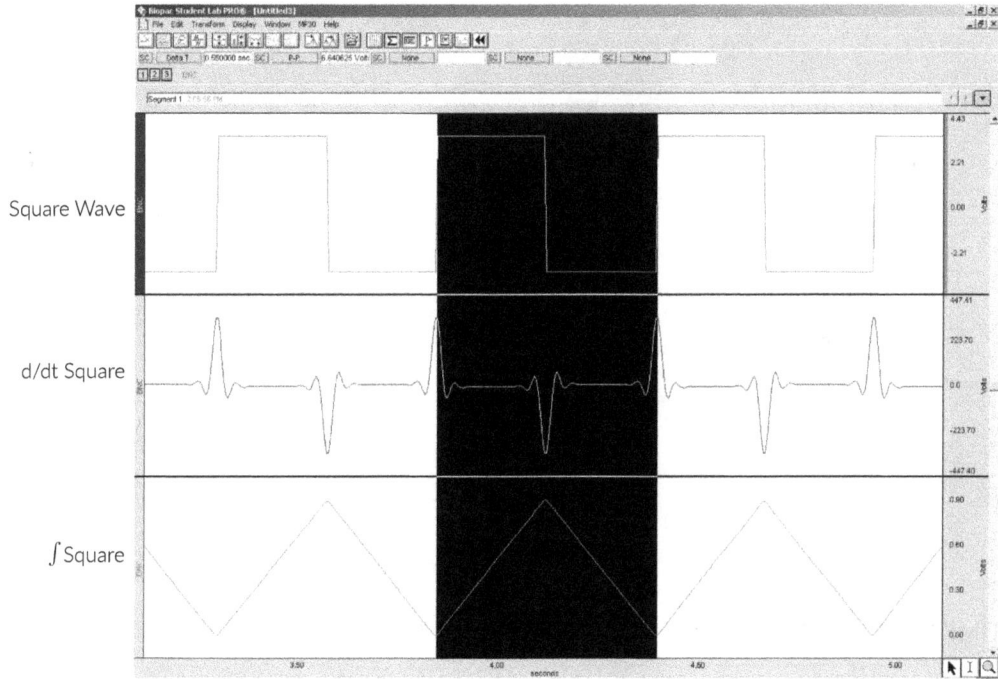

Square Wave

Gain	X200
Acquisition rate	200 (Hz)
Period	0.500 (s)
Frequency	2.0 (Hz)
Vpp square	6.641 (v)
Vpp d/dt square	671.107 (v)
Vpp ∫ square	0.896 (v)

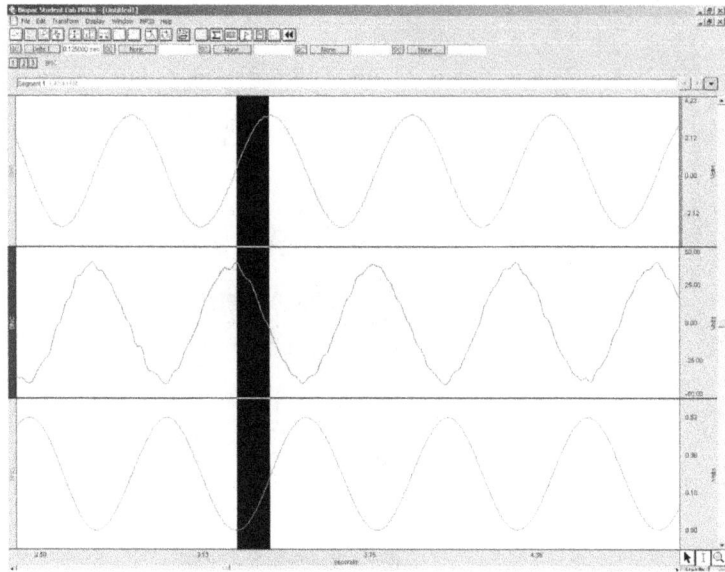

TIME SHIFT: CHANGE IN TIME BETWEEN PEAK OF D/DT SINE AND PEAK OF SINE.

Δts	0.125 (s)

Simulink

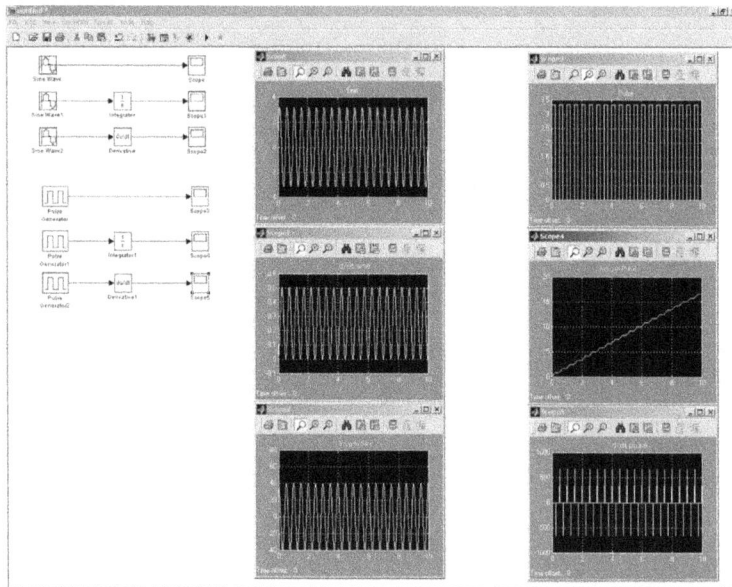

Figure Credits

LAB 3

Time Signals Design

OBJECTIVE

In this experiment, we will generate and record a few important time-signals that are useful for system tests and modeling biological waveforms. Also, in this case, the waveform will be specified, and the student will adjust the instruments to provide the desired wave. The wave will then be recorded to verify.

KEY LEARNING SKILLS

- Sine wave offset
- Bias
- DC offset
- Step function
- Impulse function

INTRODUCTION

The Sine Wave

Due to the fact that many biological systems possess pacemaker activity they will often display time data that is periodic and oscillatory in time. The sine wave is a convenient model for oscillatory biological system data. Since it is more usual that the oscillation occurs relative to a steady DC level, this has been added to our sine voltage below.

The basic sine function is represented mathematically as:

$$A \sin(\omega t + \phi) + DC$$

$$\omega = 2\pi f$$

Where:

A = the amplitude of the sine wave
phi = the phase of the signal
ω = the angular frequency
f = the frequency in Hz
DC = steady voltage level

A sample sine wave is shown in Figure 3.1 for the above relation:
Let A = 1, f = 1 Hz, and phi = 0 and DC = 0.5

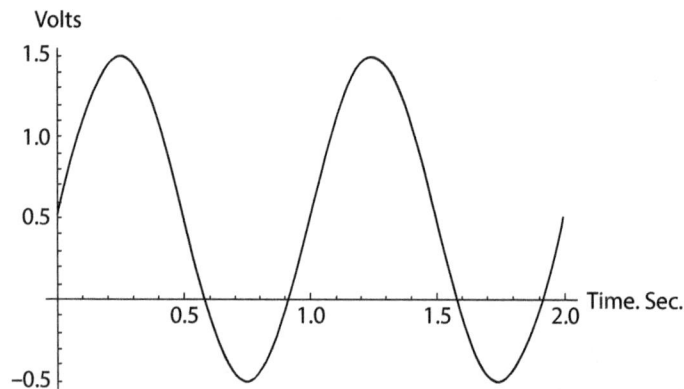

FIGURE 3.1 Sample Sine Function of Time and DC.

The Unit Step Function

In some cases, we desire only to perturb the system briefly while observing a response. There is a simple way of examining the time behavior of such a system. In this class of time signals, we begin with the step function. The step function is simply defined as a signal to be a value of one, beginning at a predefined time; otherwise it is 0.

Figure 3.2 illustrates a step function beginning at time t = 1. Since the amplitude is one, we refer to this waveform as the unit step. Note that this unit step is a theoretical waveform because it is discontinuous at time t = 1.

Physically you will find that it is not practical to create such a waveform in the lab. In particular, you should observe from your data that our step does not possess an infinite slope at the time of the discontinuity.

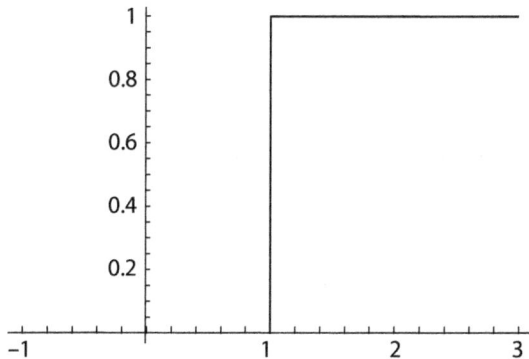

FIGURE 3.2 The Unit Step Function.

EQUIPMENT LIST

- PC with Simulink and BIOPAC BSL *PRO* software
- BIOPAC MP36 analog to digital converter system connected to PC
- Function generator
- BIOPAC BNC adaptor cable
- Clip to BNC cable
- BNC-to-BNC adaptor

BIOPAC EXPERIMENT

A. Sinusoid: In this first experiment, it is required to adjust the waveform generator to obtain specific sine wave outputs. You will then verify your results using BSL *Pro*. Calibration is very important here since you are to generate a specified voltage.

Procedure:

1. Open Biopac Student Lab Pro (BSL *PRO*).
2. Plug in the BNC clip cable into the channel 1 of the MP36 unit.
3. Calibrate channel 1 of the BIOPAC MP36 system.
4. Pull down the MP36 menu in BSL *PRO* and input the following settings below.
5. Go to setup channel and input the following data:
   a. Preset: BNC(−10V to +10V)
   b. Set/change parameters:

      | Filter = 1 | Type: None | Freq: 66.5 Hz |
      | Q = 0.5 | Gain: X200 | Offset: 0 |
      | Input coupling: DC at 1k Hz LP | | |

6. After calibration is successful, connect channel 1 BNC directly to the waveform generator output.

7. Next, from the MP36 menu proceed to setup acquisition:
   a. Record and save once using the hard disk.
   b. Sample rate = 200 samples/second.
   c. Acquisition length = 15 seconds.
8. On the function generator push in the sine wave button on the top right-hand side.
9. Set the frequency multiplier to X1.
10. Adjust the frequency such that it reads 1 Hz.
11. When you press the start button on the Biopac, you should receive a graph that looks like a sine wave. If not, make use of the vertical/horizontal autoscale buttons on the top left of the graph window to better fit the data to the screen.
12. Now change the amplitude and the period using the function generator to match the graph in Figure 3.1. Use the *Amplitude & Offset* dials to adjust waveform to give amplitude of +1.5 and −0.5 volts. It is best to set amplitude first, then the offset. Both controls affect the max and min voltages.
13. Use the window measurement cursor to fill in the table below and verify that you have the correct sine data rate: amplitude = 1/2(PP).
14. Print your final sine if it meets spec.

SINE WAVE MEASUREMENTS

	P-P	Frequency	Max	Min
CH1				

B. Step Function Generation Procedure:

Using the same set-up as the previous exercise, adjust the function generator until you can record a waveform that resembles the unit step function identified as in Figure 3.2.

Start by selecting the square wave button on the function generator. Please note that though the function generator does not have the capability of generating a step function, you can approximate this waveform by using a long period or low frequency square wave setting. Do not be concerned with starting the wave at t = 1. But make certain that the wave is 1 volt for 2 seconds or longer.

As before, print your results of the step wave record. Use the cursor tool to fill in the data in the table.

STEP WAVE MEASUREMENTS

	P-P	Frequency	Max	Min
CH1				

Using the measurement cursor, measure the slope of the rising edge of the step wave. You may find it easier to first zoom on the rising edge. This slope is defined as the "slew-rate." The slew-rate is the limitation of digital computer speed.

Go to the transforms menu and perform the derivative operation on your step wave and compare to the original step. Notice that it is limited at the peak. See sample graph at the end of this experiment for your reference.

Simulink

Reproduce the exact waves that you recorded in the BIOPAC experiment in a Simulink model. That is, the sine and step waves. MATLAB contains a built-in step block.

Group Lab Report Questions

1. Provide your Simulink and experiment results for the sine and step waveforms. Make sure that they all meet the design spec.
2. Explain how to obtain a sine wave of the same frequency as Figure 3.1, but instead with a maximum value of 3 and a minimum value of –1.
3. Perform question 2 using Simulink.
4. Discuss the problems that you had in attempting to create an ideal stepwave.
5. Prepare a written summary of your lab. Identify any relationships with your lectures.
6. Given the equation for a sine wave, does the wave generator possess the three parameters: amplitude (A), angular frequency (omega), and phase? Relate each parameter to an adjustment that is available on the wave generator.
7. How does your experimental stepwave differ from the "ideal" step function mathematical concept? Why?

Individual Lab Report Questions

1. Write a mathematical equation that represents the sine wave that you created in this experiment and let it = S(t).
2. Write the derivative of the signal S(t) above.
3. Write the integral of the signal S(t) above. Sketch your result graphically.
4. Name three engineering terms that describe the shift of the sine wave in this experiment.
5. Give an example of an electrical system signal that a step function could represent.
6. What are the features of a step signal that make it a useful system test signal?

Sample Lab Data

Sine with DC = 0

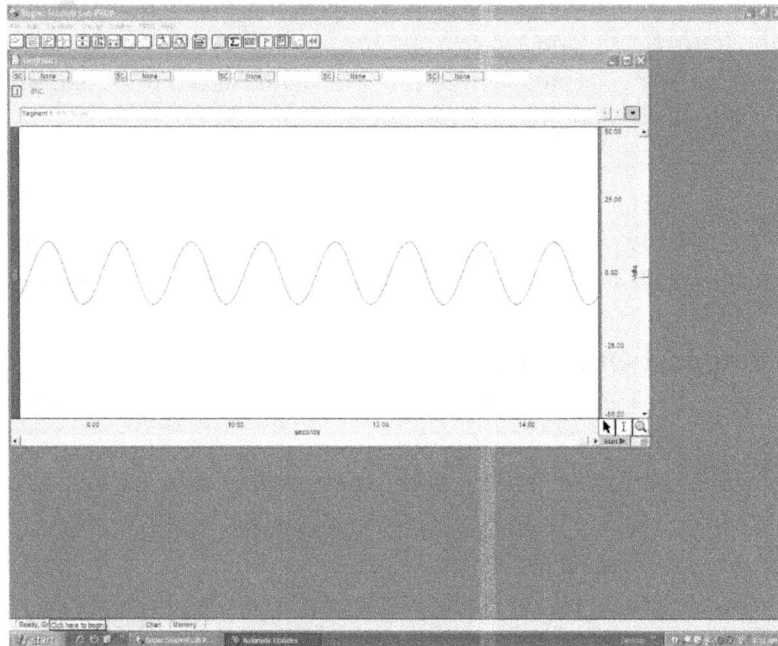

Sine with DC = +0.5

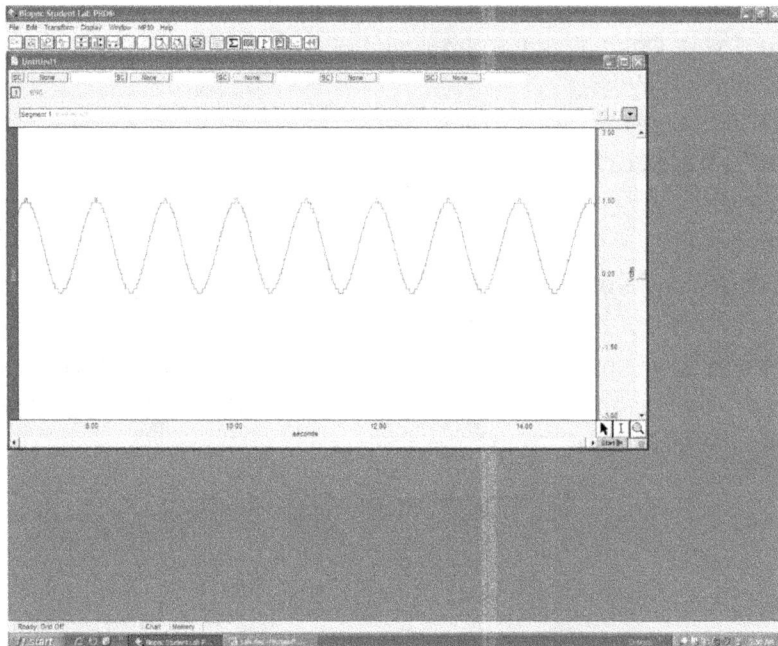

Simulink

Sine with DC = ±0.5

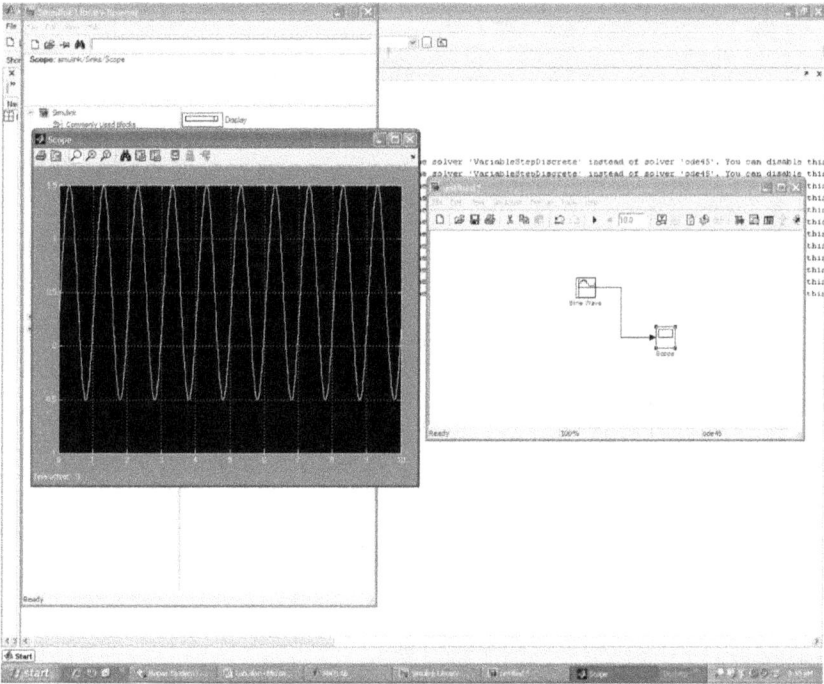

Sample Unit Step Function

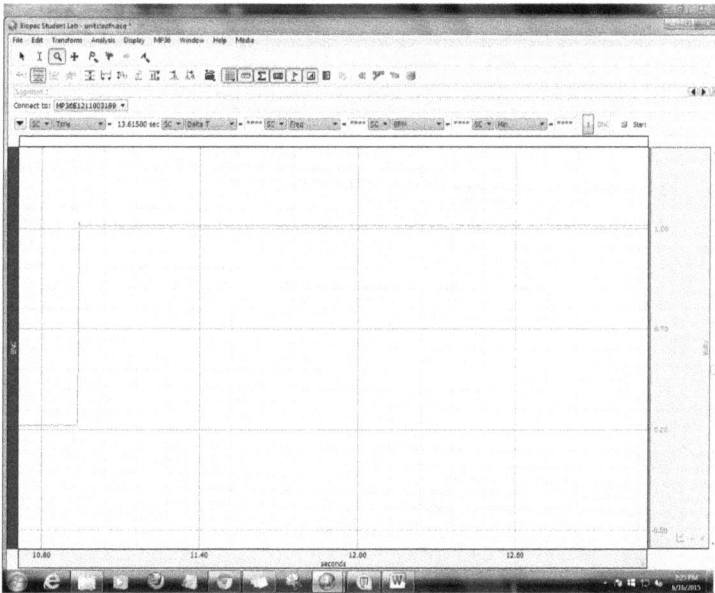

Sample Unit Step Function Derivative

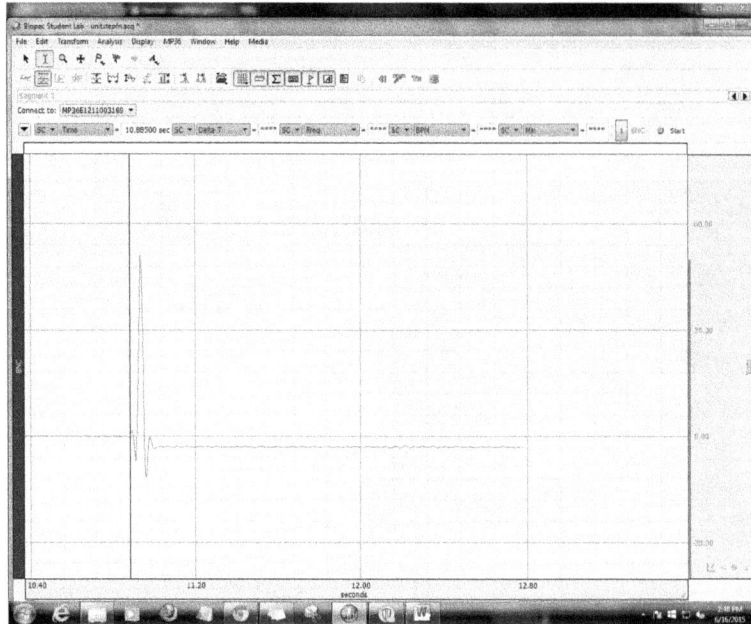

Figure Credits

Time Signals for Physiological Modeling

OBJECTIVE

In this experiment, we will record a blood pressure pulse waveform and attempt to model the pulse with a sine wave from the wave generator. Key parameters of the sine wave will be altered to design the proper waveform to match. This activity is basic to Fourier analysis.

KEY LEARNING SKILLS

- Signals
- Arterial pulse waveform
- Pulse plethysmograph (PPG)
- Modeling
- Skill correlation

INTRODUCTION

Modeling of Physiological Waveforms

Calibrations are not necessary in this experiment. Choose one partner to wear the PPG for pulse recording while another adjusts the instruments for best results.

EQUIPMENT LIST

- BIOPAC system and Simulink
- BNC to Biopac cable
- Wave generator
- PPG pulse sensor

BIOPAC

Connect the PPG to channel 2 of the BIOPAC system. Wrap the pulse transducer around the tip of your index finger. The transducer should fit snugly, but not to the point where your blood circulation is impaired. Begin with a lighter fit and increase tightness if data recording is poor.

From the setup channels menu check the three option boxes and select PPG (0.5–35 Hz) preset.

Channel 1 will record the output of the sine waveform generator while channel 2 will record the pulse (PPG) sensor.

For the simplest set up, it is best to verify that the PPG channel and wave generator channels are working separately before a recording of both is performed.

You will want to adjust the function generator output to model the physiological signals as best as possible. Also, adjust sampling rate and other parameters as in earlier experiments to obtain the best data. Sample graphs are attached at the end of this lab.

After you have verified that you can record a sine wave and a pulse wave separately, set BSL *PRO* to record both channels simultaneously.

You will find it easier to have one lab partner adjusting the frequency and amplitude of the sine-generator to match the other's pulse wave. The best way to start will be to first measure the pulse frequency and (P-P) amplitude, then set the wave generator frequency to the same values.

Please note that the best results for the pulse wave come when the patient is relaxed, still, and breathing steadily. Make sure that the lab partner providing the pulse is aware of this. Joking around will introduce large errors in your data.

When you obtain your best result, print your data and record the key waveform measures (p-p, max, min, frequency) for both the pulse and sine wave.

Lastly, place the BIOPAC window into the scope display mode by clicking the scope button on the top of the window. This will overlap the pulse and sine wave for your last data quality check. Print this result. Exit the scope mode and then use your measurement tools to find the pulse period for five heartbeats. Save that data in your notes. Create a table of beat-to-beat measurements.

Correlation Experiment

In this experiment we use the BSL *PRO* software to approximate the correlation between a sine wave signal and your arterial pulse.

Calculate the math correlation between the CH 1—Sine ("CH" channel 1) and the Ch 2—Pulse ("CH" channel 2), as you learned in your lecture theory. Refer to the "cross-correlation integral." The BSL *PRO* software does this automatically for you, but in this case, we will do the calculation explicitly using the math transform feature. The purpose is so that you can observe the effect of the correlation calculation on the waveforms, step-by-step.

Correlation of the Sine and PULSE Waveforms

1. From the menu bar go to **Transform** > **Waveform** Math.
2. Select **CH1 * CH2** and choose **NEW** (this will allow a new plot to be generated).
3. Once the new (3rd) plot is generated, go to **Display** > **Auto scale Waveforms**.
4. From the menu choose **Transform** > **Integrate**, check the remove mean box, then click OK.

To identify if the sine and pulse waveforms are correlated:

- If plot is *greater than 1*, correlation has occurred between the two waves.
- If plot is 0, there is *no correlation*.
- If plot is *negative*, there is an *inverse correlation* relationship between the two signals and retesting must occur to produce the desired positive correlation response. Print your final window.

SIMULINK

You will model the pulse data with a single sine wave. Do not import the pulse into the Simulink. Pay attention to key timing data points and max and min values from your pulse recordings. For example, pay attention to the correct frequency, wave shape, and 0 crossing times. You will be evaluated based on the average accuracy of your Simulink model's representation of your acquired waveforms.

Group Lab Report Questions

1. Provide your Simulink results for the sine waveforms and all BSL *PRO* graphs.
2. Discuss problems that you had in attempting to model the pulse with a sine wave.
3. Calculate the average heart rate for your subject over five heartbeats. By what percentage did the heart rate increase or decrease for each of the five beats?
4. Is the sine wave generator a good average model of the pulse waveform? Discuss.
5. Prepare a written summary of your lab and discuss its relevance to your lecture materials.
6. Discuss your correlation results.
7. Explain why the PPG sensor is an indirect measurement of blood pressure.

Individual Lab Report Questions

1. In this experiment, you determined the first harmonic of the Fourier series of the pulse. Describe the procedure that you would use to find the second and third harmonic sine wave amplitude and frequency.
2. The correlation values in this experiment were not normalized to be within the −1 to +1 range. How would you modify your correlation calculation to normalize its value?
3. Describe the extra calculations that you would perform on the BIOPAC system to complete question two above.

Sample Lab Data

Time Signal for Physiological Modeling

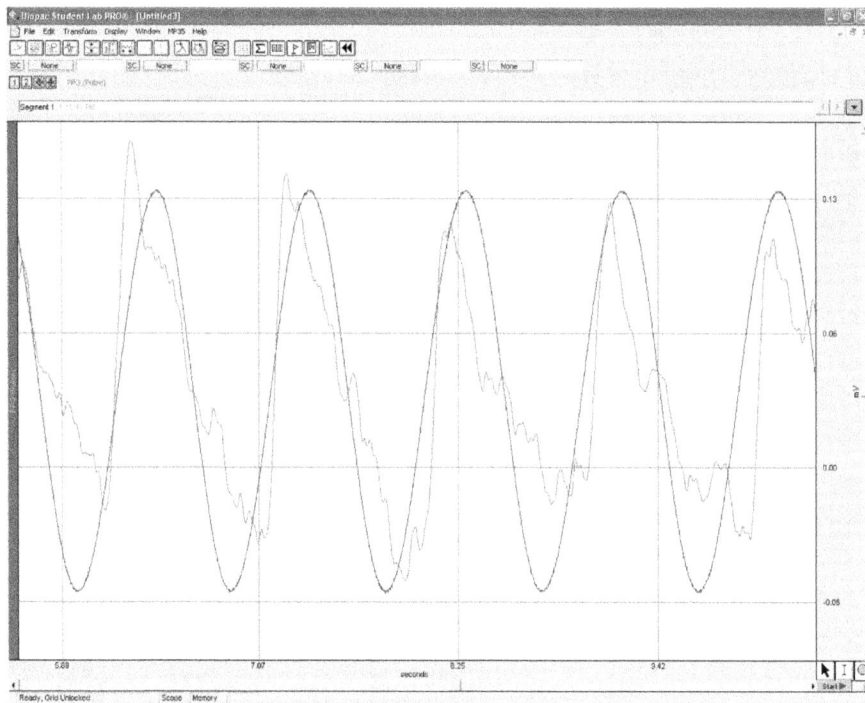

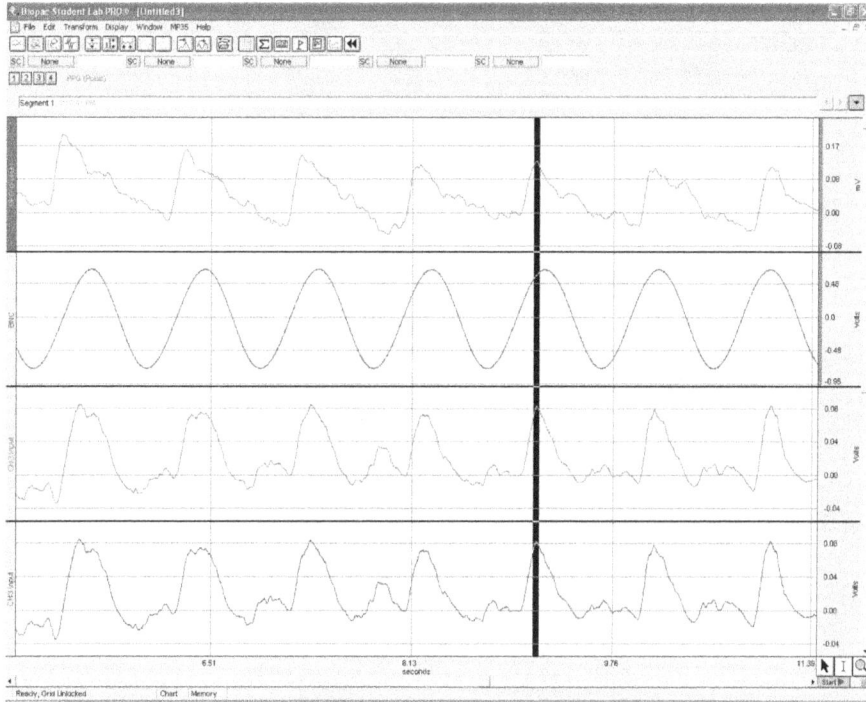

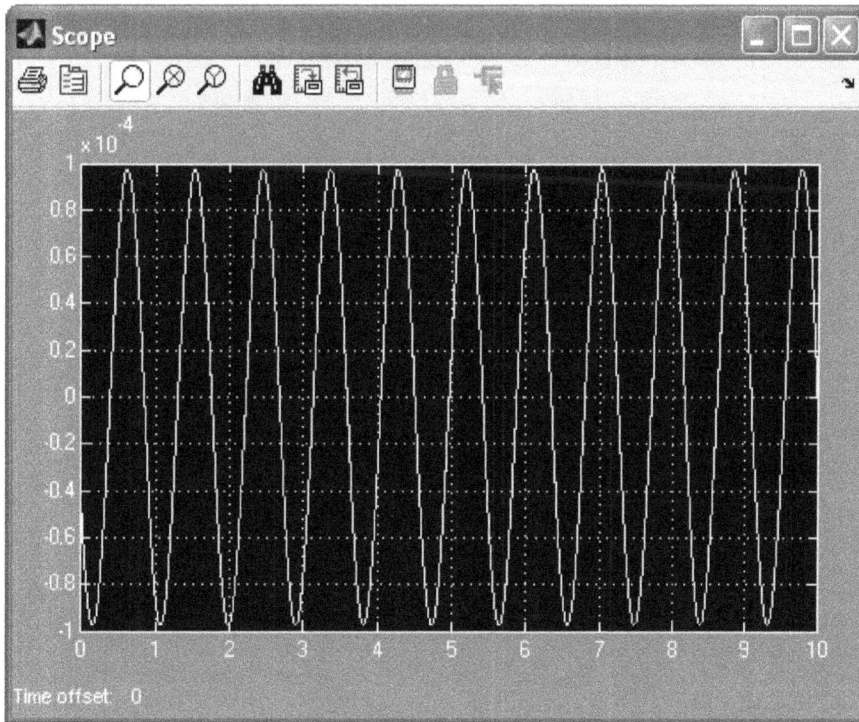

Figure Credits

LAB 5

Noise and Variability

OBJECTIVE

In this experiment, we will begin by generating and recording a sine wave's time signal as in the previous lab. In this case, however, we will alter the voltage source and the BIOPAC amplifiers. The purpose of these changes is to introduce "noise" into our recordings. When noise is present in our work, as in common day experience, the recording conditions are not perfect. The signal recorded is then in error.

KEY LEARNING SKILLS

- Types of noise
- Characterizing noise signal properties
- Noise reduction
- AC power
- Random noise
- Moving average
- Low pass filter

EQUIPMENT LIST

- PC with Matlab and BIOPAC system
- Waveform generator
- BNC to BIOPAC cable
- BNC to clip lead cable
- BNC T-connectors (2)
- RC Box

INTRODUCTION

Noise may be viewed as any signal that interferes with our ability to measure signals, biological or otherwise. Signal processing techniques permit the noise to be reduced or sometimes eliminated; the amount of contaminated data will fall greatly. When recording biological signals, noise is often caused by the recording environment. It is often best to first attempt to improve your data recording by removing the noise source. Signal processing is effective in restoring the data to its original form, but it should be used very carefully; signal processing itself can contribute to data contamination.

Noise/Random Noise

Power system noise sources of noise can result from electric fields, coupled with AC power lines. This type of noise is easily recognized: it will always be at the same frequency as the power line, usually 60 or 120 Hz.

Unfortunately, in many cases, sources of noise are not straightforward and often unknown. In our case, a noise model is used. In particular, it is most common to employ a random noise model, for example, either a random uniform noise or gaussian noise. These noise models are usually preferred because they are well known mathematically. Moreover, these noise models are most commonly employed in the derivation of signal processing techniques. Although a respectable start in improving a data record, unfortunately most kinds of observed physical or biological noise do not follow the characteristics of gaussian noise.

Gaussian noise remains the most commonly employed approximation by researchers. We will examine its features, particularly in the time domain. Gaussian noise assumes that the occurrence of an "event," x is governed by the gaussian probability function modeled by function p(x) and graphed in Figure 5.1.

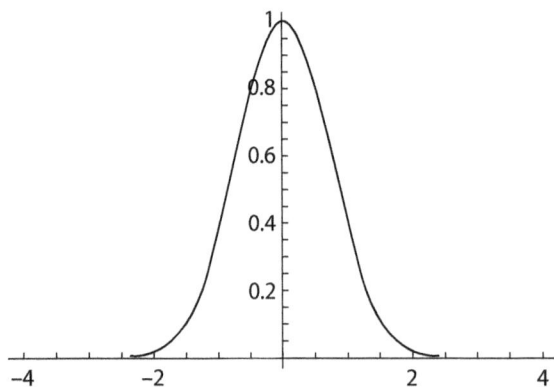

FIGURE 5.1 Gaussian probability function P(x).

If we replace x by time in the probability function, we can then use P (x(t)) to generate a time series of random numbers whose probability is governed by P(x). This is defined as the Gaussian-random time series and is shown below in Figure 5.2.

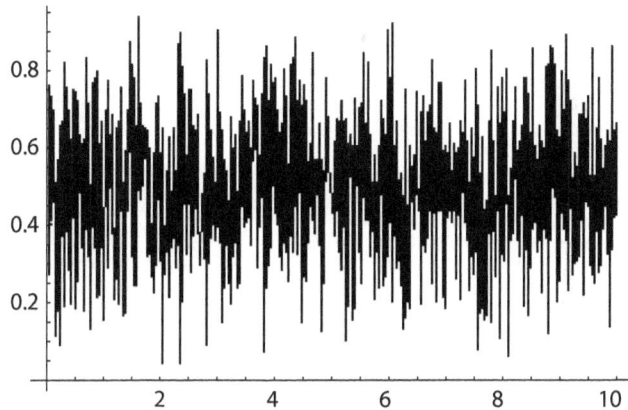

FIGURE 5.2 Gaussian-random time series waveform approximation.

BIOPAC

A. Noise Experiment (60 Hz power system noise)

Use two clip leads from the BNC cables to connect the waveform generator to a channel input of the BIOPAC MP36. Place a 1 MΩ in series with the red clip leads following Figure 5.3. The black ground clips are connected directly together as ground.

Begin by setting up the channel input for the BNC voltage preset on channels 1 and 2. Set the gain for both channels to be X200. Adjust the waveform generator to provide medium voltage output in the free-run mode.

Dial in a frequency of 1 Hz. Set the data acquisition rate to be 100 Hz and the recording duration for 15 seconds. Select store to memory.

Push the button on the waveform generator to select square wave generation. Start the recording, and you will observe and measure a square wave signal on channel 1 and channel 2. Now, increase the sample rate to 1000 Hz. Start another recording, but this time one group member should touch the end of the resistor indicated in Figure 5.1 (as touch point). Notice the recorded signal is not a clear square when the wire is touched, resulting in the top and bottom of the wave becoming noisy. Touch the resistor a few times to record a few noisy square waves. You can compare the noisy square wave to that recorded in channel 1, which does not possess the series resistor.

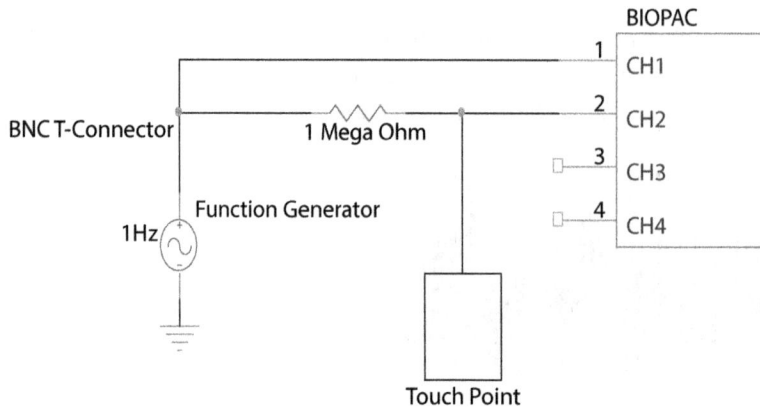

FIGURE 5.3 Connection diagram for noise experiment.

At this point you should have a clear square wave on channel 1 and a noisy square wave on channel 2 (see sample data at the end of this lab).

Examine the distorted top and bottom via the zoom tool. Using the measurement cursor, find the frequency of the noise. This frequency is likely to be some multiple of 60 Hz. If not, repeat your recording while touching the 1 MΩ resistor. But make certain that your subject is standing to make sure he/she is not grounded. It will be best to use someone with rubber-soled shoes to prevent current to the floor.

Use the measurement cursor to also find the peak-to-peak voltage of the square wave and then (P-P) voltage of the 60 Hz noise. Write down these values to use in the Simulink model later.

Now attempt signal processing and improve the recording on channel 2 by using a digital filter: select channel 2 then go to the transform menu and select digital filters IIR low pass. In the filter window, select the Hanning window and leave the numbers of coefficients at its default value. Check the box to filter the entire waveform.

Do not choose to display and print the filter response function.

Compare the filtered channel 2 to the clean channel 1 square wave. Observe whether or not the filter has restored the noisy square wave back to the original square wave signal. If the noise is not yet removed, adjust the filter cutoff frequency lower and try again.

B. Electronic Noise

Again, record the square wave on channel 2, but now change the channel gain to 5000X and set the resistor for 100 kΩ so that there is a lower resistive connection to channel 2. Acquisition rate should be 1000 Hz. Record again and compare channel 1 to channel 2. Again, use the magnify tool to observe the details of the noise on channel 2. Use the measurement tool to check the noise frequency and (P-P) voltage. Record your measurements. Apply signal processing to improve the recording on channel 2.

Select the channel 2 waveform and go to the transforms menu to find smoothing. This operation simply averages the adjacent number of points that you choose together and replaces the data with the average of the data. Since we know that the square wave is 1 Hz, the top and bottom of the wave should be about 1000 points. Choose a 100-point smoothing transform and apply it to the entire noisy recording on the waveform.

Now go back to the channel 2 set-up and restore the gain to 200X and acquisition rate to 100 Hz. Do another recording and you should find that channels 1 and 2 have nearly identical shapes. Observe any differences if possible.

SIMULINK

Design a Simulink model for the case of the direct connected square wave and the resistance connected square wave source.

Assume that the electrical noise produced by touching the resistor is additive in your model.

Construct a second Simulink model to represent the high-gain electronics noise condition. Again, assume that noise is additive and use the white noise block as the model noise source.

Print your results for both models after you have adjusted the parameters to model your BIOPAC data. Take care to model the same S/N ratio as your data.

Group Lab Report Questions

1. Provide copies of the following graphs:
    a. 60 Hz noise
    b. Amplifier high-gain noise
    c. Simulink (a+b)
    d. High-gain noise with smoothing filter
2. In the case where you touched the resistor, what were the characteristics of the noise that you observed? For example, what was its frequency? What was the signal-to-noise ratio?
3. What is the origin of the noise that you observed? Draw a complete circuit.
4. Did the low-pass filter remove the noise?
5. After low-pass filtering, was the square wave restored to its original shape as in channel 1?
6. Draw the voltage source circuit for channel 1 and channel 2 separately.
7. What were the characteristics of the recording with high-channel gain? Be quantitative. Report your measurements.

8. What is the source of the noise in the high-channel gain condition?
9. Draw some basic conclusions about the best approach to obtain noise-free data. What would you do in the future to improve the input circuit after studying question 6?
10. Provide a written summary of your lab.

Individual Lab Report Questions

1. Design an RC filter that could be placed at the channel 1 input of MP36 in place of the digital filter that was used. Provide your circuit diagram and circuit element value calculations.
2. Using your 60 Hz noise voltage data calculate the amount of 60 Hz noise current that passes through your subject's body.
3. Provide a circuit diagram that you used for your analysis. (Hint: assume that the input current to the BIOPAC channel input is negligible.)
4. Referring to the setup circuit Figure 5.3, explain why the channel 2 noise is always larger than channel 1.

Sample Lab Data

Noise and Variability

POWER SYSTEM NOISE

1 Vpp Square wave and Square wave w/ introduced Noise through 1 mOhm

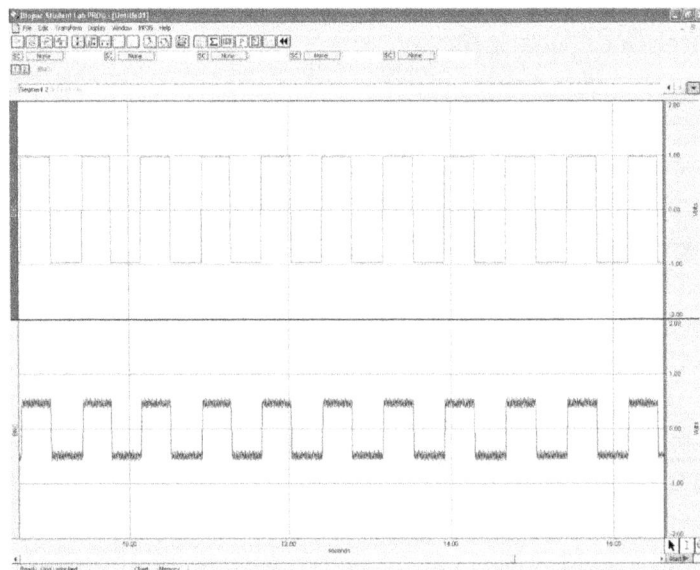

60 Hz Noise

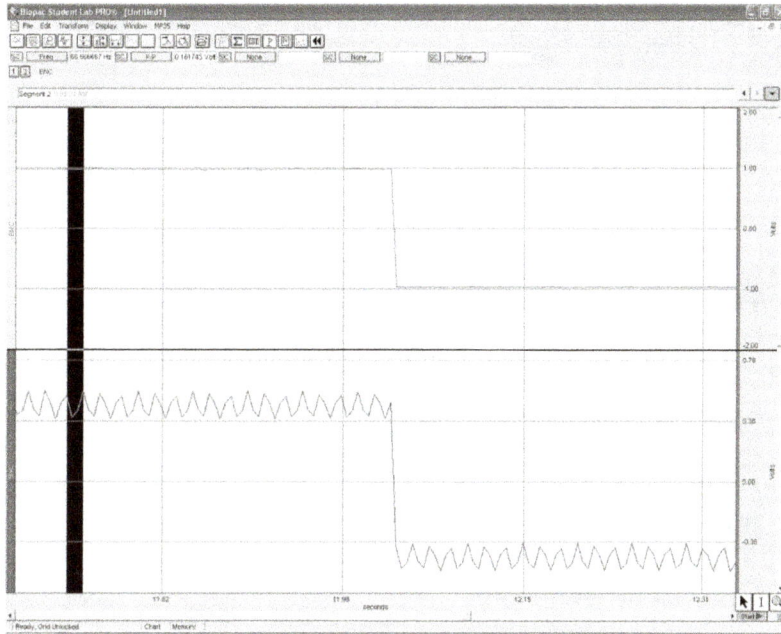

Applied Low-Pass Filter with fc = 30 Hz

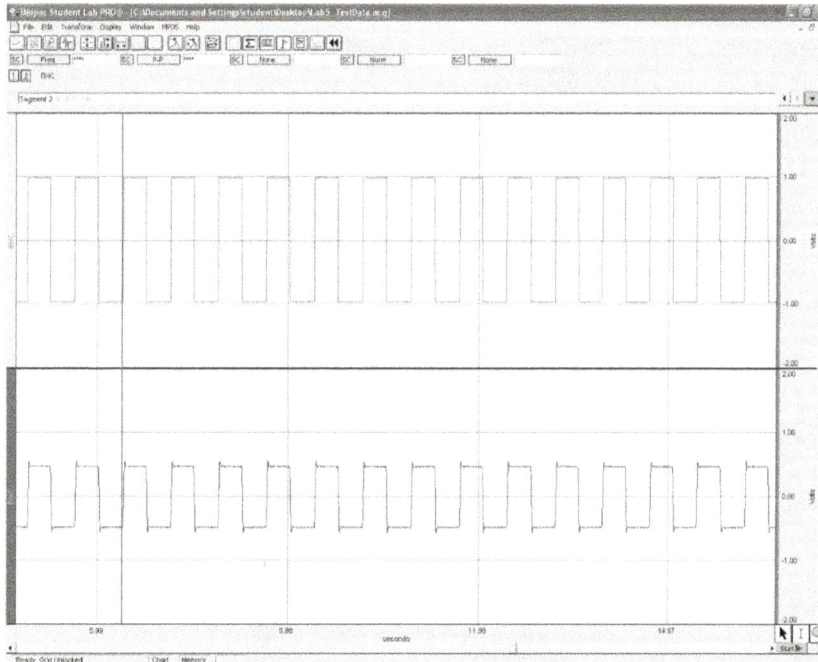

Electronic Noise

100 K electronic noise

100-point Applied Smoothing

SIMULINK

60 Hz Noise

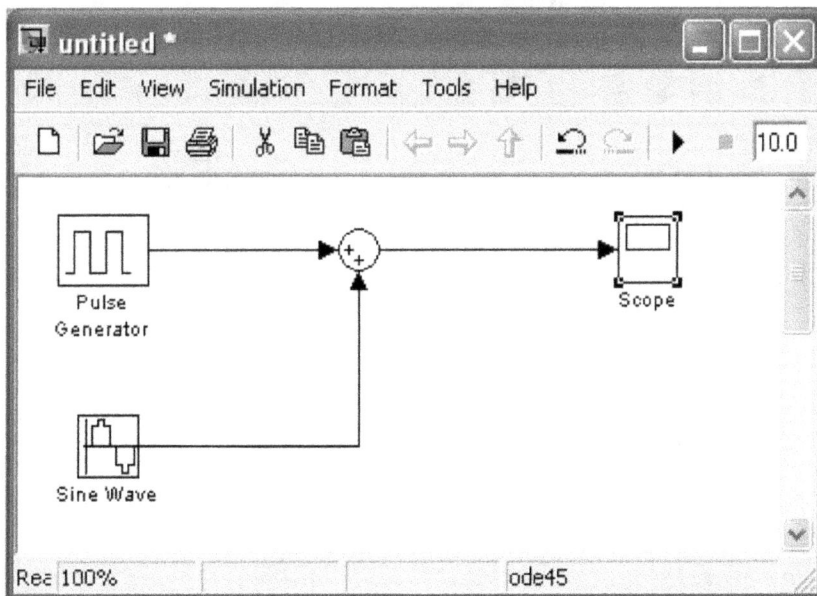

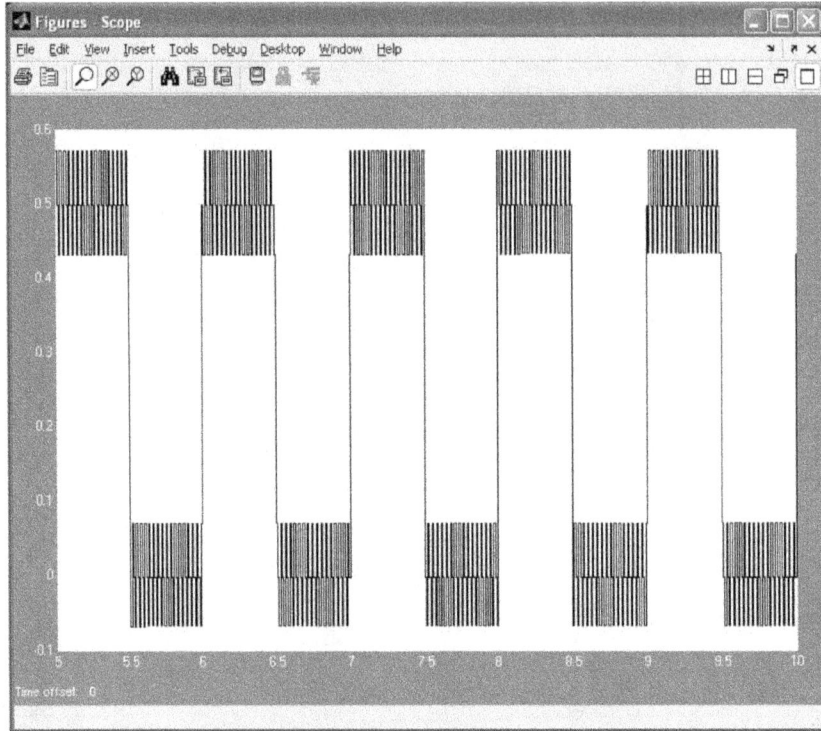

Bandlimited White Noise

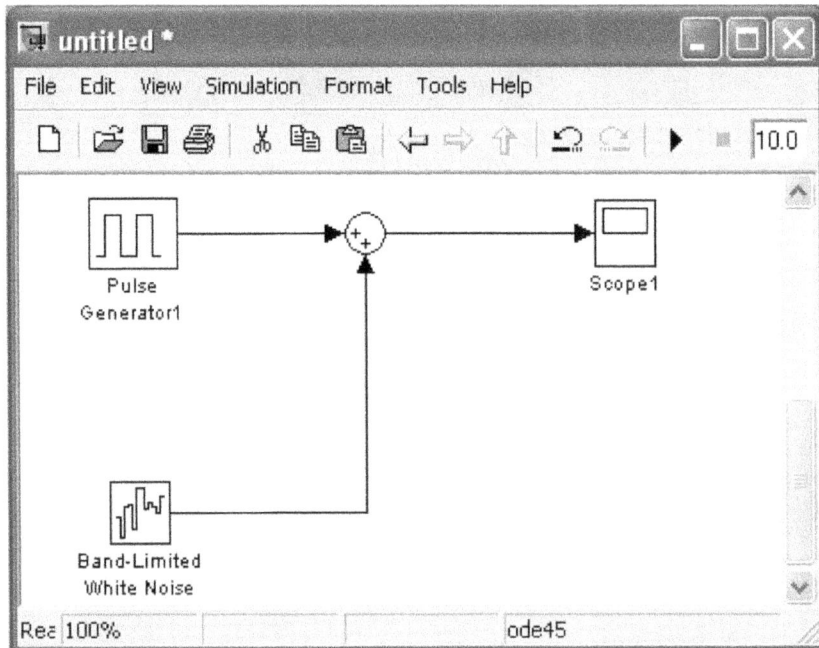

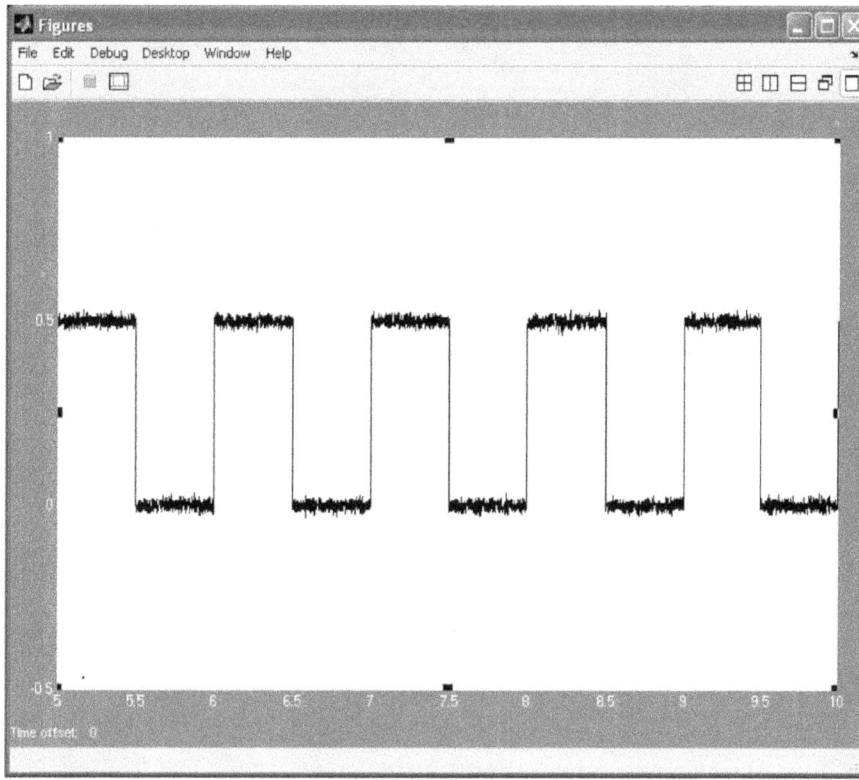

Figure Credits

Part II

Systems

LAB 6

Fourier Transform

OBJECTIVE
To record a physiological waveform and create a model using Fourier analysis.

KEY LEARNING SKILLS
- Fourier modeling
- Inverse Fourier transform
- Waveform synthesis
- Frequency Spectrum
- Magnitude and phase

INTRODUCTION

Modeling Physiological Time Signals
In a previous lab, time signal for physiological modeling, we recorded an arterial finger pulse. As part of that experiment we also attempted to match a sine waveform to the recorded waveforms by trial and error; adjusting the time waveform generator until a reasonable visual match was obtained.

As expected, a perfect match could not be obtained. However, although the sine wave match was not perfect, the error present was not unreasonable. We performed other standard wave measurements on the pulse wave such as frequency, amplitude, and correlation. This experiment suggested that a sinusoidal time signal was a good choice in modeling the arterial pulse waveform.

Conceptually, this is the reasoning behind the application of the Fourier transform. The Fourier transform renders a time signal into a frequency continuum of sinusoidal amplitudes and phase. The resulting sinusoidal continuum then exactly matches the original time signal.

Recall the Fourier integral of a time signal f(t):

$$f(j\omega) = \int_{-\infty}^{+\infty} f(t)e^{-j\omega t}\,dt$$

Since we will be working with real-world data, it is not possible to use this calculus version of the Fourier transform. Instead we will use its discrete form.

This permits us to integrate the data time series that is sampled by the BIOPAC system. Specifically, the BIOPAC system supplies a numerical algorithm that is called the fast Fourier transform.

The Square Pulse

While this lab focuses on the transform of the physiological arterial pulse, it is also useful to explore some common signal waveforms at this time.

One such man-made waveform is the square pulse. The electronic pulse that will be examined here is the digital TTL pulse that is important to digital and computer systems. A sample pulse is shown below in Figure 6.1.

FIGURE 6.1 TTL digital Pulse Waveform example.

In this experiment, the interest is in the frequency analysis of waveforms. Referring to Figure 6.1, the wave can immediately be observed to be periodic and therefore possess a base frequency. Additionally, observe that the wave contains very rapid changes from a low to high voltage. In your classroom examples of Fourier transform it was found that rapid changes in a waveform would require a relatively large proportion of high frequencies to model the wave as a Fourier series. Remember this concept when you examine the data of this experiment. Because the pulse contains frequencies that extend from the base frequency on up to several multiples of the base frequency, the pulse wave is said to possess a broad frequency spectrum content. This information is valuable to a systems engineer. For example, since it is known that the pulse contains a wide range of frequencies it may be used as a system test input to evaluate its transfer function over the same large range of frequencies. You will do this later in experiment 8.

In chapter 5, noise was examined from two different kinds of sources.

Additionally, examine the Fourier transform of the random noise and explore its frequency spectrum. The Fourier transform of noise was provided in Figure 1.6 (Semmlow 2005). Reviewing that figure, you may recall that pure random noise possesses a spectrum that is fairly constant (uniform) across many frequencies over a wide range. It is then said to possess broadband frequencies. In this lab you will test this calculation using recorded data in the Biopac system similar to your lab 1.

BIOPAC EXPERIMENT

EQUIPMENT LIST
- PC with BIOPAC system and software
- Simulink software
- BIOPAC pulse sensor (PPG)

Frequency Spectrum of Arterial Pulse
Procedure: (Pulse)

Start by connecting the BIOPAC pulse sensor to the channel 1 input of the system.

Choose a member of your group to be the subject and apply the PPG sensor to his/her index finger. When applying the sensor, be sure to have a correct fit, with a snug but not tight fit. Setting the fit to be too tight will cut off blood flow and obliterate the pulse signal.

Now set the BSL *PRO* software to the pulse sensor (PPG) preset from the BIOPAC menu.

Do a test recording of the pulse and verify a good pulse waveform. If the pulse sensor appears to be noisy, position the sensor to a different finger and try again.

If the data still do not show a good pulse record, choose another member. Do this until you are able to obtain a satisfactory pulse signal. Sample recordings are shown at the end of this lab.

Use the BSL *PRO* software measurement tool to determine the period and heart rate of your pulse data (beat-to-beat for several beats).

Select the entire waveform using the cursor, then go to the transforms menu and choose Fourier. Choose a Hanning window and select the pad with last points. Select linear amplitude instead of log. Plot the phase result along with the amplitude. Remember to select remove mean.

These choices direct the Fourier algorithm to make the pulse data continuous beyond the beginning and end of your recorded data; these selections essentially make the data record to appear as a continuous periodic waveform. If we did not make these choices, the Fourier analyzer would be performing its calculation on a waveform that is not what we assume to be analyzing. For example, if the endpoints of the waveform are not set equal to the first and last data points (padding), the Fourier transform would assume a small step function at the end points.

Your resulting transform will appear in a new window. Notice that the axis is now changed to frequency instead of time.

Use the cursor measurement tool on the Fourier window to find the peak magnitude, phase, and frequency at three frequencies. Record these data in a table for later use.

SIMULINK

Now, referring to the Fourier analysis data that you obtained in your BIOPAC experiment, you will design a Simulink model that will output the pulse data that you just recorded.

The approach that we will use is to approximate the Fourier transform with a summation of sine functions (Fourier series). Refer to your Fourier transform graph; you will find that some frequencies possess a large magnitude. Thus, if we choose only those frequencies that have these dominant peaks, when we add them together, most of the frequency components of the pulse should be represented in your Simulink model. We can mathematically represent the pulse time signal as the sum of sine functions as follows:

We have chosen only three sine functions to reconstruct the pulse signal. This is an approximation, since your FFT generated 100 sine points to be accurate.

Sample FFT is supplied at the end of this lab section.

Observe that the magnitude FFT transform contains at least three dominant peaks. As a general rule, peaks less than one half of the main peak are too small to be included in the reconstruction model. In this example we can read off the magnitude of each peak directly from the graph. They each correspond to the values of A_1, A_2, and A_3.

The phase can also be read from the graph for each corresponding peak. Peak 2 is 112 degrees.

Construct a data table for frequency, magnitude, and phase. Phase must be converted to radians for the final Simulink model and results in the following Figure 6.2:

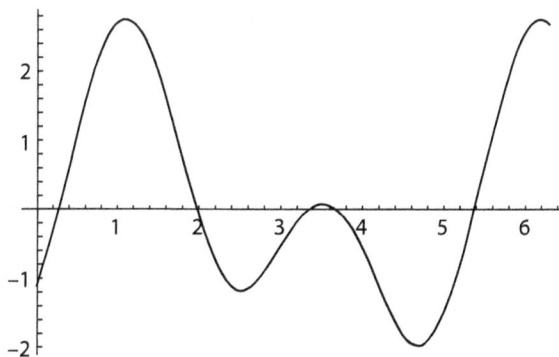

FIGURE 6.2 Fourier model of recorded arterial pulse data synthesis.

FREQUENCY SPECTRUM OF NOISE

Record baseline noise in BSL *PRO* following the same procedure in the lab 1 of this lab manual. Set the channel gain to x5000. After a good baseline noise recording is obtained, go to the analysis menu at the top and apply the FFT function. A screen will pop up with different options. Use pad with 0s, check dB, magnitude and for window use Hamming. Press OK. The new graph of the Fourier transform will appear. The frequency spectrum should be uniform.

FREQUENCY SPECTRUM OF A SQUARE PULSE

To record a square pulse into BSL *PRO*, connect the BNC adapter cable to channel 1 of the MP36 and connect the other end to the TTL of the waveform generator. Make sure the generator frequency is set to 1 Hz. Next, set up channels and use BNC, +/−10 volts for the preset. Use a gain of x200. Next, set up acquisition. Save once using memory and set the sample rate to 1000 Hz. Use an acquisition length of 20 seconds. Now start to record the pulse. After the square pulse is recorded, apply an FFT to your recording as you did earlier. You should notice that the graph is a constant magnitude uniform spectrum except at low frequency where you see the fundamental 1 Hz pulse frequency as a large magnitude. Sample graphs are provided at the end of this lab to check your results.

Group Lab Report Questions

1. Provide a copy of all graphs: pulse, FFT mag and phase, Simulink pulse model.
2. Find the error between the heart rate determined from the pulse-time data versus the FFT frequency of the primary amplitude.
3. Measure the pulse period for every pulse recorded. Are they the same? Find the mean period for all pulses and calculate the mean heart rate from this.
4. Do an error analysis of your Fourier pulse model. Discuss your sources of error. How would you improve your results?
5. What do the first and second frequency peaks on the frequency graph represent physiologically? What physiology causes heart rate to speed up and slow down?
6. If the period of each pulse was not constant, why?
7. Prepare a written summary of your experiment.
8. Do you observe Fourier frequencies less than the heart rate? What do they represent physiologically?
9. What other physiological waveforms can be modeled as in this lab? What signals would be a poor choice for Fourier synthesis?

Individual Lab Report Questions

1. Show that the measured FFT frequencies are equal to or near integer multiples of the fundamental frequency.
2. Was your FFT fundamental frequency equal to the mean heart rate? Provide an error analysis.
3. What was the bandwidth of the pulse waveform that you recorded? Do not include noise or sinus arrhythmia in your bandwidth.
4. Compare the first term of the FFT with that measured in lab 4 ("Pulse modeling with correlation"). You should compare the frequency and amplitude of both sinusoids.
5. Provide an error analysis to show how closely they agree.

Sample Lab Data

Pulse and Fourier Transform

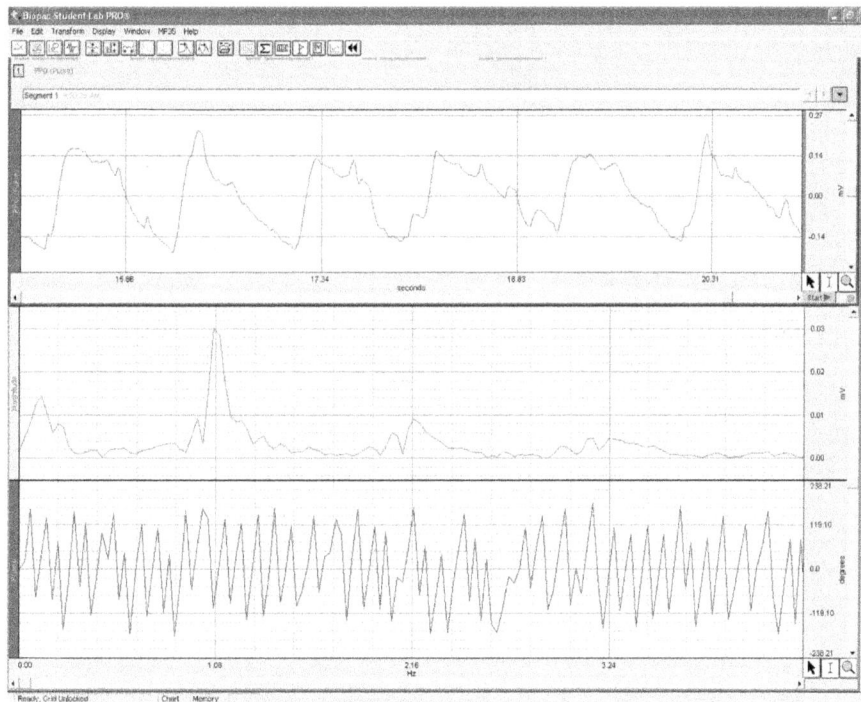

Magnitude

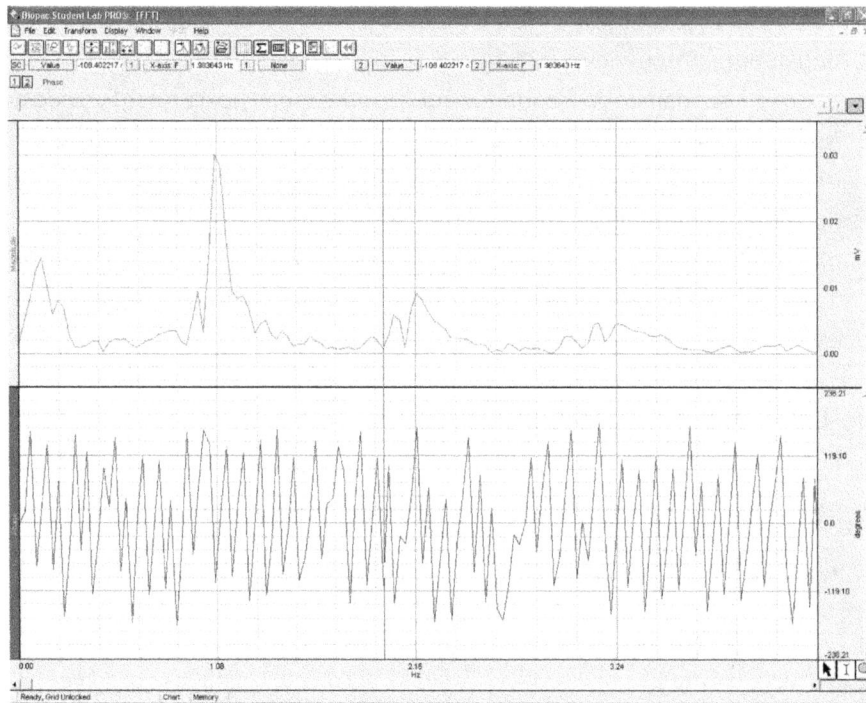

SIMULINK

Phase

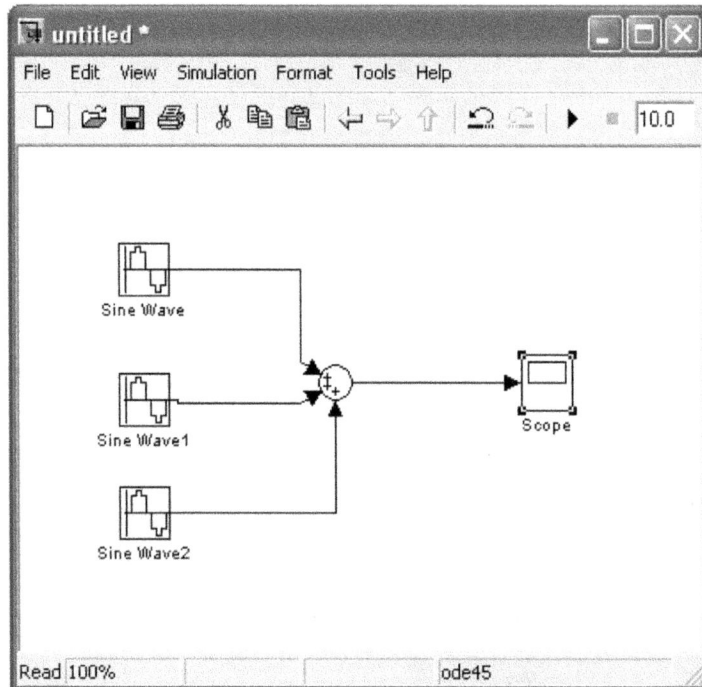

Fourier series of pulse

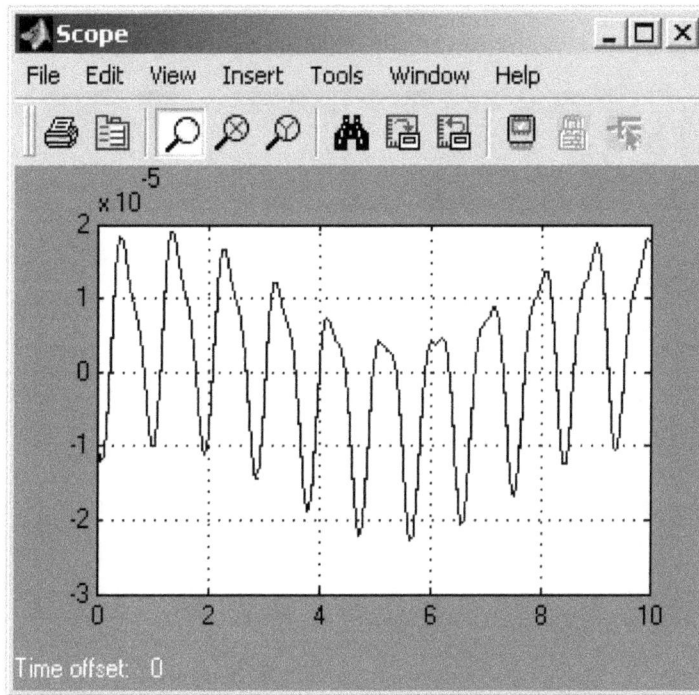

Fourier transform of square pulse magnitude spectra

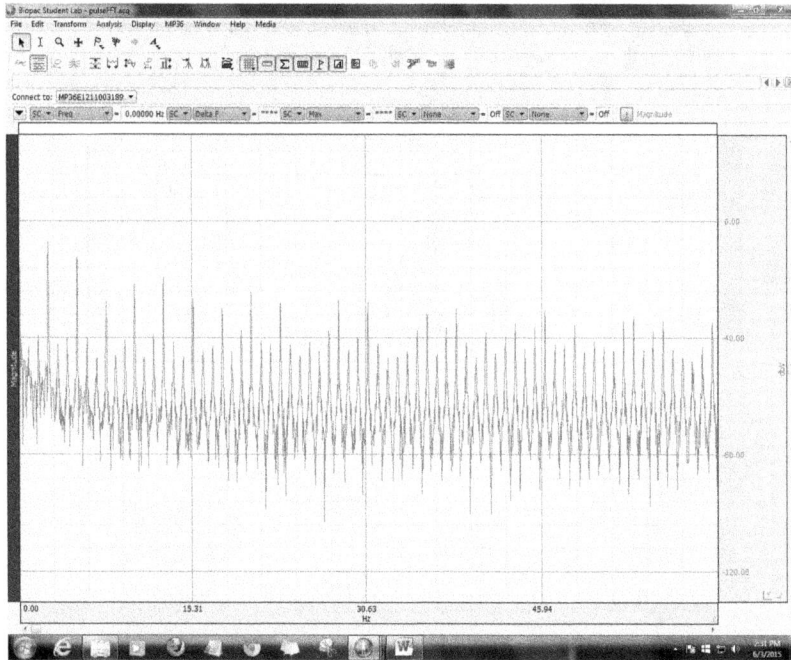

Sample square pulse record

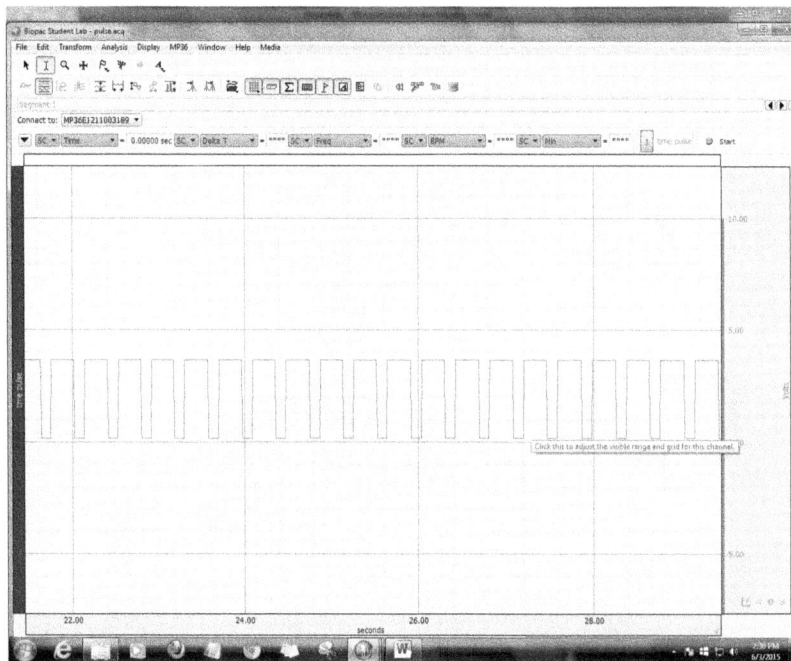

Sample FFT of BSL *PRO* Noise recording. Line indicates uniform Frequency spectrum

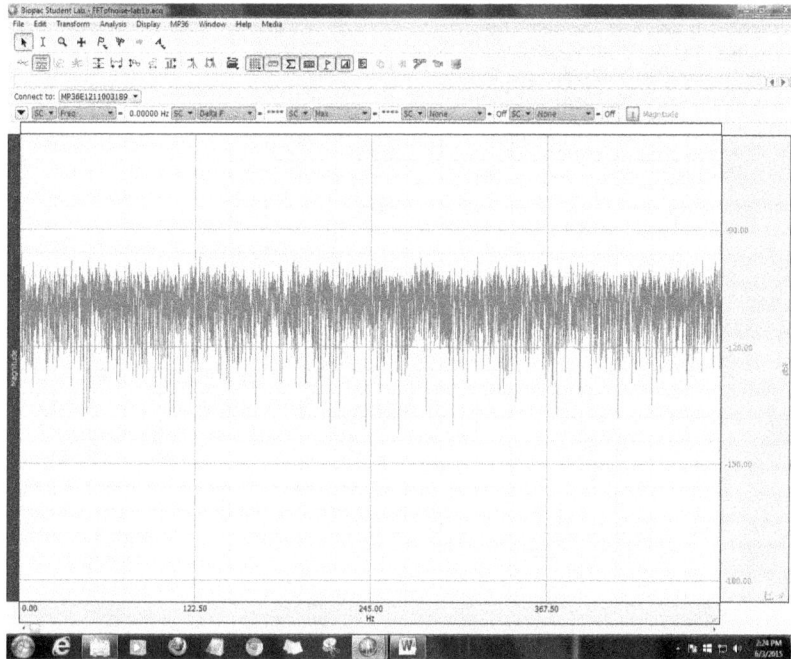

Sample BSL *PRO* Noise Recording

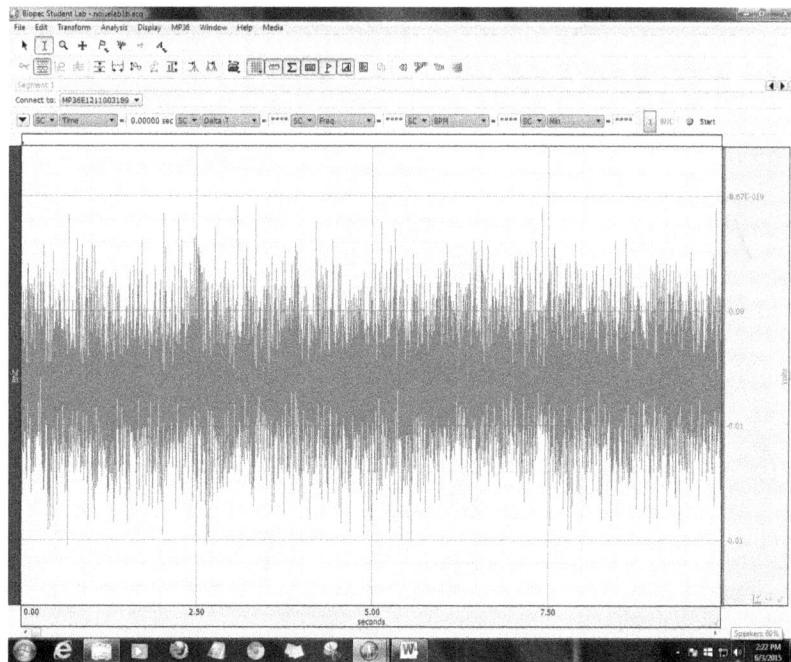

Figure Credits

LAB 7

Time/Frequency (Transfer Function)

OBJECTIVE

To examine the relationship between time and frequency domain in one biomedical system where Fourier will permit us to move from frequency to time domain and vice versa.

KEY LEARNING SKILLS

- Transfer function
- Frequency plot
- Impulse response
- Sine response
- Cutoff frequency
- Bandwidth

INTRODUCTION

A. Time/Frequency System Response

So far, we have examined systems by their response to our basic set of time signals for testing. For example, we introduced the step function and the sine function. Recall that the Fourier transform is the following.

$$f(j\omega) = \int\limits_{-\infty}^{+\infty} f(t)e^{-jwt}\, dt$$

Furthermore, the systems that we have studied have been linear. As a result of this, input and output of the system are directly proportionate.

Take the condition that we input a step function, u(t), to a system. Now we can record the system response to the step as a signal, y(t), to the step response. If we find the derivative of the system response, dy(t)/dt, we can say that this is the response of the system to the derivative of the step function. But, by definition, du(t)/dt = δ(t). So, we can conclude that the derivative of the system step response is the impulse response of a system.

But the impulse response is the system transfer function h(t). Now going back to the Fourier domain, let the input to a system be X(jω). In this case we have X being the impulse, whose Fourier transform is 1.

Thus, in our linear system, we have Y(jω) = H(jω)*X(jω). But since the input x is the impulse, we find that the impulse response of a system is the transfer function of the system, and equivalently the derivative of the step response is the transfer function of the system, h(t), in the time domain

$$dy(t)/dt => H(t) = \text{transfer function[time domain]}$$

$$\textbf{\textit{F}}\{h(t)\} => \textbf{\textit{H}}(jω) = \text{Transfer function I[Frequency domain]}$$

We can see the value in having the derivative of the step response, y(t) of a system, because its Fourier transform is the system transfer function.

Recall from your lecture theory, the transfer function tells us everything that we need to know about how the system will react to any signal. But, be careful, because it works only for linear systems and many biological systems are not linear.

EQUIPMENT LIST
- PC with BIOPAC system and software
- Simulink software
- BIOPAC stethoscope sensor
- Miniature loudspeaker
- Function generator
- BNC to clip leads cable

BIOPAC EXPERIMENT
Stethoscope Transfer Function by Means of the Step and Sinusoid Response Methods.

A. Pulse Response Procedure:

- Start by connecting the BIOPAC stethoscope to the channel 1 input of the system. We will be testing the transfer function of this sensor for sound frequencies.
- Set the BSL *PRO* software to the following settings:

Stethoscope Setting for the BIOPAC System

Open the channel menu and select the stethoscope. The channel presets for the stethoscope will not need to be altered. Choose the preset heart sound.

Go to the data acquisition menu and set the sample rate to be 1000 Hz. Set the recording duration to 15 seconds. Save to hard drive. Save once.

Now start a recording in the BSL *PRO* data window. Lightly tap the stethoscope bell to verify that you are recording a signal. Start another recording, but this time apply a brief but sharp tap with your fingernail to the stethoscope. This tap will simulate the step function input to the stethoscope system.

You should see a sharp pulse on the recording that oscillates briefly, but quickly dies out. Select your tap recording with the cursor. Go to the math operations menu and select the differencing operation.

This operation is doing an approximation to the derivative for us as follows, and is accurate for only very small time differences.

Measure the oscillating frequency and record it in your notes.

In our case, the acquisition rate was set to 1000 Hz, so our step is 1 ms and is adequate for measuring acoustic frequencies. Once you have performed the differencing operation on the stethoscope recording, you should select the pulse with the cursor and select the Fourier transform. Select to plot magnitude and phase graphs. Now you have the transfer function of the stethoscope. Print these graphs.

B. Stethoscope Frequency Response Transfer Function by the Sine Wave Response Method

We obtained the transfer function for the stethoscope sensor from the derivative of its step response in the time domain. From this response we performed a Fourier transform to obtain the transfer function in the frequency domain.

Now we will obtain the transfer function directly by applying sinusoidal sound at separate constant frequencies to the stethoscope sensor and record responses over a range of many audio frequencies (to a sine wave source of sound). We are assuming that the sound amplitude and phase are constant.

Connect a BNC clip lead cable to the output of the function generator. Connect the clip leads to the speaker terminals. Leave the stethoscope connected to one channel.

Start up the function generator and put it in the free-run mode.

Set the starting frequency to be 50 Hz.

Adjust the output until the speaker provides audible sound; too much output, however, will cause the speaker to distort the sine wave. Place the stethoscope bell in front of the speaker but be sure not to make contact. Keep a piece of foam rubber in between the speaker and scope.

It may be necessary for one lab partner to hold the speaker so that it does not vibrate on the table.

Set the BSL *PRO* to start a recording. Verify that you can record a clear sine wave. Measure the (peak-to-peak) value of your sine wave. Set the acquisition duration to 15 seconds. Start another recording. This time change the frequency dial from 50 Hz to 60 Hz. Measure the (peak-to-peak) again. Construct a table in your notes of frequency and (P-P) values.

Gradually continue this process of sweeping the frequency higher and higher by increments of 10 Hz. You may need to adjust/increase the acquisition rate to record a clear sine wave as you increase frequency. Switch to higher frequency multipliers and continue to sweep higher in frequency until the stethoscope provides little output. Now do the reverse: do the same while sweeping the frequency lower below 50 Hz. Do not change the function generator output amplitude at any time during the sweep. As you get down to 20 Hz you will find that the output becomes noise and there is no longer a measurable sine. At this point you are finished and should have a table of (P-P) values for each frequency input of 50, 60, 70, 80, and 100 Hz.

Note the frequencies where the sine wave amplitude decreases to 0.707 of the maximum overall (P-P). These are above and below the max and are the cut-off frequencies. Calculate them from your table.

Measure the stethoscope (P-P) amplitude at the frequencies. Also, the stethoscope may not permit you to measure frequency above 100 Hz since its output will be too low. You may enter your data directly into Excel for plotting. This is the transfer function from the Sinusoidal input method. There is no Simulink model for this lab.

Follow the procedure that you used in the earlier labs to make certain that you are maintaining quality sine waves. Record the sine amplitude measurements at each separate frequency in a table for your report. Do not attempt to measure the phase here.

Group Lab Report Questions
1. Submit FFT mag graph.
2. Plot the sine wave amplitude versus frequency for the stethoscope.
3. Compare your sinusoidal amplitude-frequency graph to the transfer function obtained via the step response and Fourier transform. Provide an error analysis.
4. Discuss the importance of the 0.707 amplitude frequency. Provide your high and low .707 frequencies.
5. Using the .707 frequencies, calculate the bandwidth of the stethoscope. Propose a Simulink model for the system. Do this by comparing your transfer function graph data to find similar transfer function examples you may find in your textbook or lectures. Suggest a mathematical transfer function based on your analysis.

6. At what frequency does the stethoscope provide the largest sine amplitude? How well does this compare to the step oscillating frequency in %?
7. Provide a written summary of your experiment, related to your lecture theory.
8. Why does the stethoscope resonate?

Individual Report Questions

1. The ideal input signal to test a system transfer function should be uniform magnitude at every frequency. We used a short tapping pulse and a blowing sound as test inputs. Discuss the frequency magnitude characteristics of these two signals and how valid they were for finding the stethoscope transfer function.
2. Look at the transfer function of the stethoscope that you tested. What type of filter best describes a stethoscope? Why?
3. What frequencies did you find that the stethoscope attenuates?
4. Physicians normally use the stethoscope to listen to the sounds of the heart valves. What is the frequency spectrum of a normal heart valvular sound?
5. Is the stethoscope transfer function you tested good for listening to the valve sounds?

Sample Graphs
Step Function Input Response
Lower wave is the differenced response

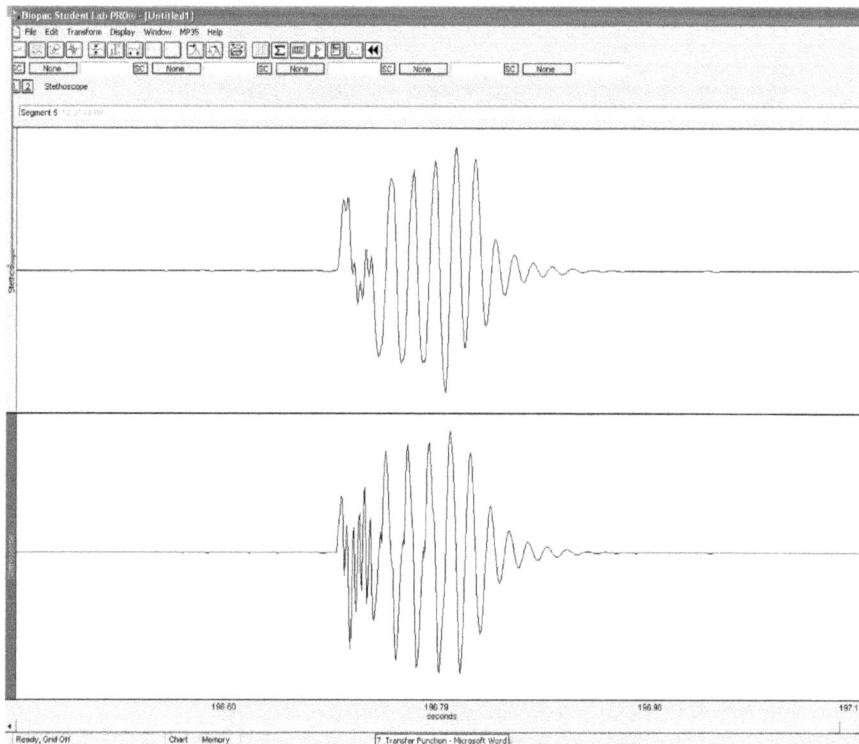

Differenced Response
FFT Transfer Function
Resonance Frequency ~52.73 Hz

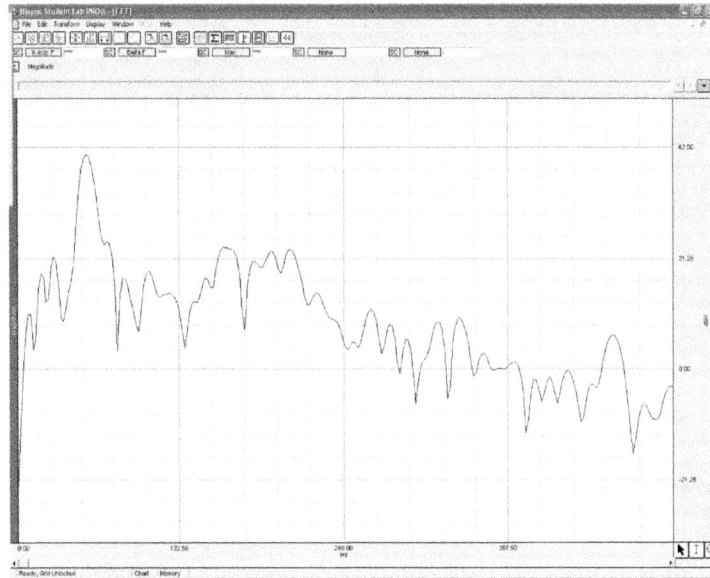

Transfer Function of the Stethoscope
Resonance frequency ~52.73 Hz

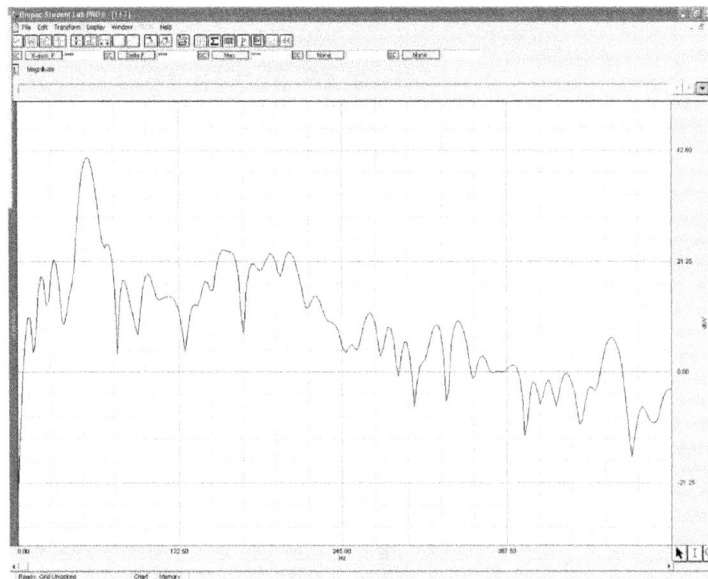

Recording of Simulated Heart Murmur [Turbulence Random sound]

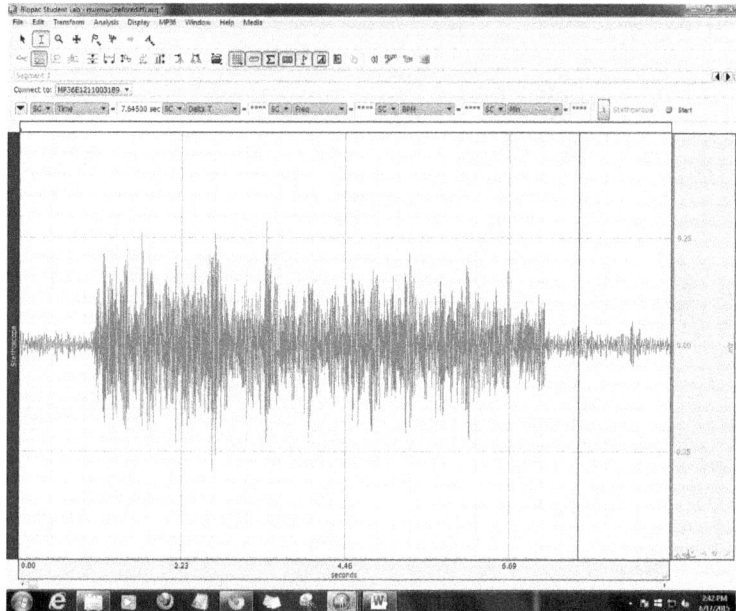

FFT of Murmur. Transfer Function of Stethoscope (assuming a random uniform noise input, compare to the step response recording)

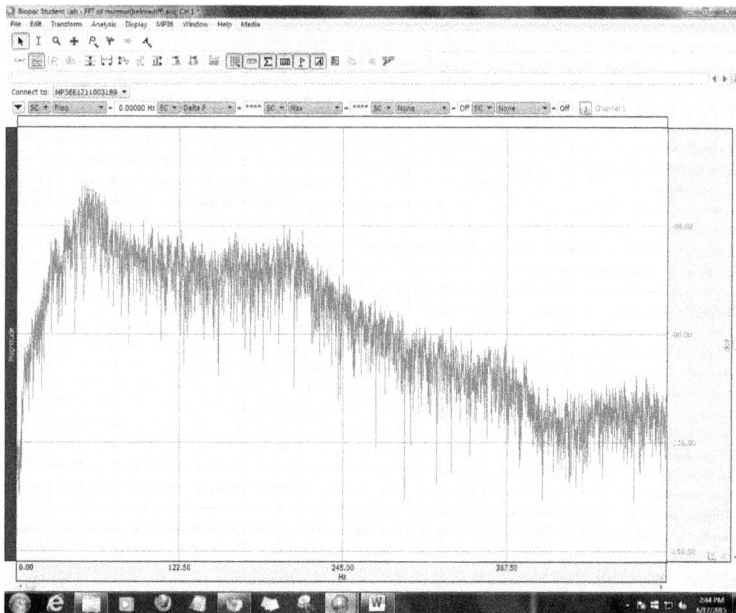

Figure Credits

LAB 8

System Modeling

OBJECTIVES

To measure the response of a SINGLE compartment, physical system in the time domain. Measured parameters will then be used again in MATLAB-SIMULINK to create a compartment model of the same physical system. The key parameters of the system will be determined by a time constant method.

KEY LEARNING SKILLS

- Compartment model
- Time constant
- Flow resistance
- Compliance

INTRODUCTION

The compartment model has great applicability in physiological modeling. We define a compartment as any enclosed volume within the body, without regard to the content of the volume V.

We can define compartments after various structural levels of anatomy. For example, a cardiac myocyte could be modeled as a compartment, as well as the cardiac ventricle, the lung, the urinary bladder, and so on. Take in point the lung; a biomedical device that could be similar to the human lung is the occlusive arm cuff, which is routinely used to measure blood pressure.

We will use this arm cuff as an air-filled compartment that we can examine in this laboratory. The physical system is shown below in Figure 8.1. As previously noted, we define a compartment as any enclosed volume within the body, without regard to the content of the volume V human lung is the occlusive arm cuff.

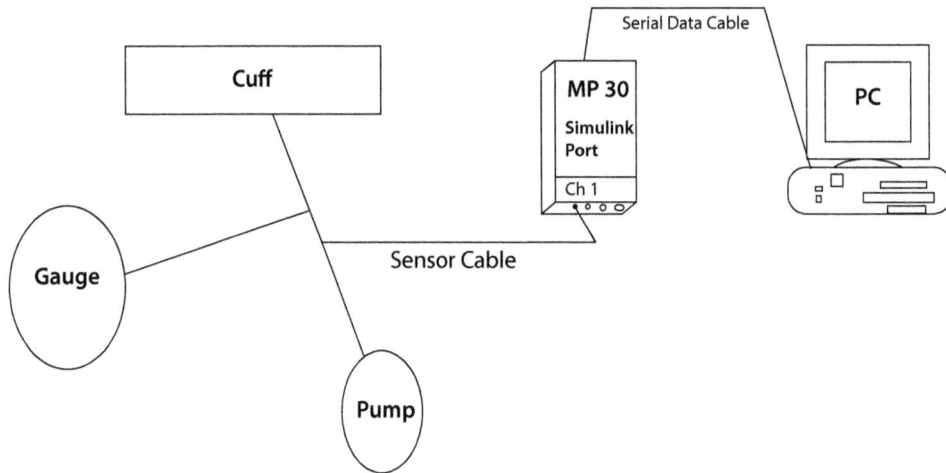

FIGURE 8.1 Occlusive arm-cuff single-compartment system.

In this experiment, the air-filled arm cuff represents the single compartment of the system.

To begin the analysis of the system, let's assume that this compartment can physically be modeled as a mechanical compliance C, where C=, and where P is the compartment pressure.

P and V are therefore the key variables for this system.

The system shown assumes that the cuff is deflated. Airflow out of the cuff is permitted through the release valve. In this case, assume that the air flow is simply proportionate to the pressure applied to the valve. That is also the compartment or cuff pressure.

Mathematically, this situation is written as $Q = P/R$, where R is introduced as a parameter and constant of proportionality. Note that this relationship indicates that air flow will be linearly related to pressure. Most likely the needle valve associated with the hand pump does not follow such a simple law. But for now, we will take it as an approximation that can be tested later, if necessary.

The flow of air out of the cuff or compartment results in a time-dependent change in the compartment volume.

Mathematically, this is expressed as:

$$\frac{dV}{Dp}$$

, and substituting the valve relationship for flow, we have,

$$\frac{dV}{dt} = -Q$$

Then dividing by C, the compliance and using the chain rule, the final differential equation results as:

$$\frac{dP}{dt} = \frac{-P}{RC}$$

Since the compartment parameters are in a product, let us redefine them to be the new constant τ = RC, such that,

$$\frac{dV}{dt} = -\frac{P}{R}$$

This now completes the derivation of a single linear compartment model.

Students will find wide application for this model in their studies within the biomedical field. Students who have had basic physics or differential equations will recognize τ as the parameter called the time constant. We can further recognize the model differential equation as a first-order linear differential equation, whose solution is easily verified to be

$$\frac{dP}{dt} = \frac{-P}{x}$$

, where P(0) is the initial condition of the starting value of compartment/cuff pressure.

Furthermore, we require that for a stable solution.

EQUIPMENT LIST
- PC with Matlab and BIOPAC BSL PRO software
- BIOPAC MP36 analog-to-digital converter system
- Occlusive arm cuff with gauge manometer and hand pump
- Matlab Simulink software

BIOPAC EXPERIMENT

Arm-Cuff Compartment Experiment
Set up and equip the blood pressure cuff following Figure 8.1. Turn the air release valve fully counterclockwise and completely deflate the cuff. To inflate, turn the release valve clockwise. Connect the pressure sensor plug to MP36 port, chapter 1.

To calibrate the pressure sensor, scroll down the MP36 menu, channel, and then go under view/changes and parameters (gain = X500), then scaling. Next, click on the button that says scaling; you can also follow the BIOPAC instruction pages for analog scaling attached to this lab. Note calibration here follows the same approach as you did for voltage except you are now scaling the axis to read pressure in mmHg.

As before, the scaling operation requires two pressures to rescale the axis. In previous labs we used 0 and 5 volts conversely for voltages because we had a calibrated voltage at 5; this time we need to calibrate the BIOPAC system so that it follows the pressure gauge reading, which reads in mmHg.

Since the gauge has already been calibrated, we will use it as the calibration source. This time it is arbitrary which pressures we pick to do the calibration. In practice it is always better to calibrate over a wide range of values to give the best precision, so here we will choose to calibrate first at 0 then at 100 mmHg.

Click on CAL1 box that initially set to 0. That button resets the reading to the new 0. Make sure that the cuff is completely deflated, and the gauge pressure reads 0 mmHg.

Next, tighten the air-release valve and pump up the pressure until you read 100 mmHg on the pressure gauge. Once that is correct, set CAL2 to be 100 mmHG and then click on CAL2 in calibration to read the gauge value, which should be 100 mmHg.

Now, under MP36, scroll down to show values and check that the meter matches the reading on the gauge. Test a few values of cuff pressure to see if the MP36 reads the same as the gauge pressure. If this is correct, you have completed the pressure calibration. Now you can set your MP36 acquisition parameters listed below.

1. Open the MP36 menu and preset blood pressure cuff.
2. Go to setup channel and input the following data:
   a. Preset: blood pressure cuff
   b. View/change parameters:

Filter 1	Type: low pass	Freq: 66.5 Hz
Q: 0.5	Gain: X500	Offset: 0
Input Coupling: DC	at 1k Hz LP	

3. Set-up acquisition:

   Record and save once using the hard disk.

   Acquisition length 1 Minute

   Note: Before you start a recording, pump the cuff pressure up to 100 mmHg. Then, slightly open the air-release valve. As the cuff pressure begins to fall, start the MP36 to record the pressure-time curve. Sample rate: 00 samples/second.

4. Find the time constant t by using the following three methods:

   Time constant method: Go back to the pressure-time recording using the tile wave-forms button. Read the time at which $t = \tau$, that is when:

$$P(t) = P(0)e^{-t/x}$$

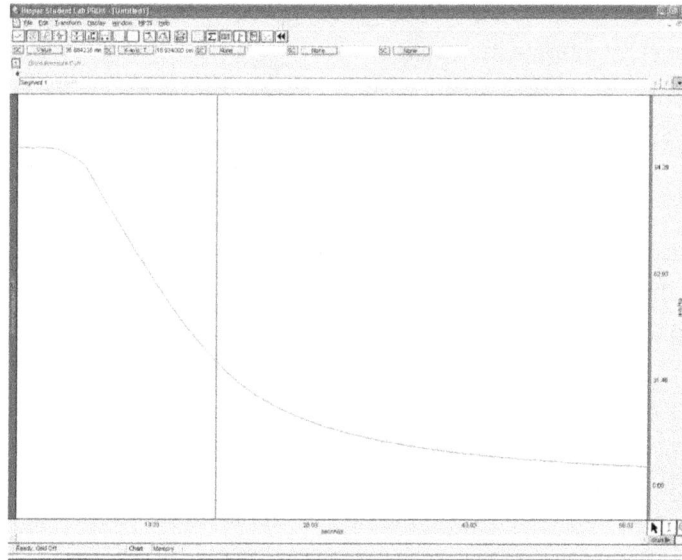

SIMULINK

In this design experiment, the Simulink functional blocks will be used to model and solve the single compartment model defined by EQ1. You may find useful the integrator, constant value, multiplier, and scope blocks in modeling EQ1.

It may be needed to manipulate EQ1. Open time constant τ = 0.5. Verify that the Simulink model is working by plotting pressure versus time and observing that the output graph reproduces the exponential solution of the differential equation provided above.

Once you are certain of your baseline result, print and then repeat the Matlab solution for three different initial conditions and time constants. In your model, set the initial compartment pressure to be 100 and input the values for three different time constants. Repeat a cuff pressure recording after slightly closing the air-release valve to provide another trial condition.

Group Lab Report Questions

1. How you would model the human lungs using the compartment modeling concept?
2. Analyze your pressure-time data from the Matlab model and the air-cuff recording to determine the time constant. Once results for the time constant are obtained, provide an error analysis and discuss how well this approach worked.
3. Submit graphs for all trials performed.
4. Provide compartment model designs for two other physiological systems.
5. Use the time constant obtained from the occlusive cuff experiment to model that experiment using the Matlab solution. Show comparison graphs for pressure-time and the phase plot graphs. Provide an error analysis. Discuss how well the Matlab compartment model was able to model the air-cuff experiment. Discuss any discrepancies

observed in any of the trials and suggest improvements that could be made to the Matlab model to improve its results.

6. The experiment that has been performed allowed the measurement of a single parameter of the system defined as the time constant RC. Design an experiment that would permit the separate determination of R and C. Can it be performed with the original apparatus?

7. The compartment model does not include the pressurization phase of the experiment; it neglects the hand pump used to fill the cuff with air. Modify and improve your Matlab model to incorporate the process of pumping. Note that physiologically this aspect of the model would be analogous to the pumping of the heart, where the aorta would represent a compliance compartment.

8. If the pump pressure were sinusoidal, what kind of filter does the air cuff behave like? Support your answer with analysis.

Individual Lab Report Questions

1. Compare the slope of the pressure curve at high pressures to that at low pressures.
2. Based on your result in question 1, sketch a plot of dP/dt versus P where P = pressure.
3. What electrical element is the cuff airbag compliance analogous with?

Sample Lab Data
Cuff pressure versus time record

Slow Time Constant

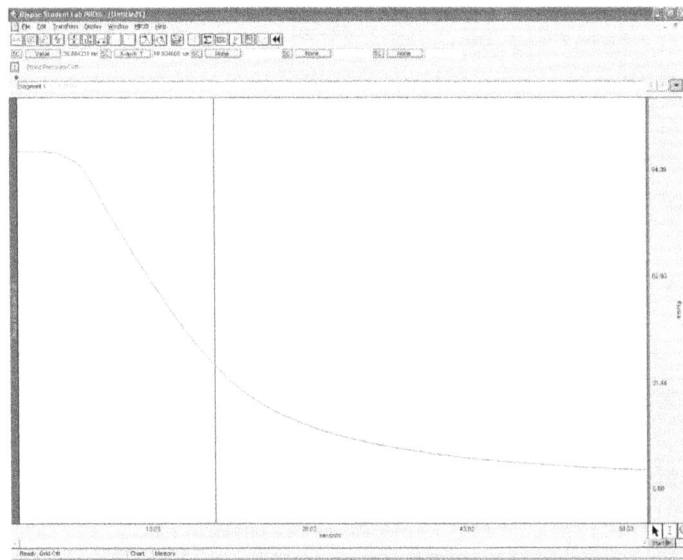

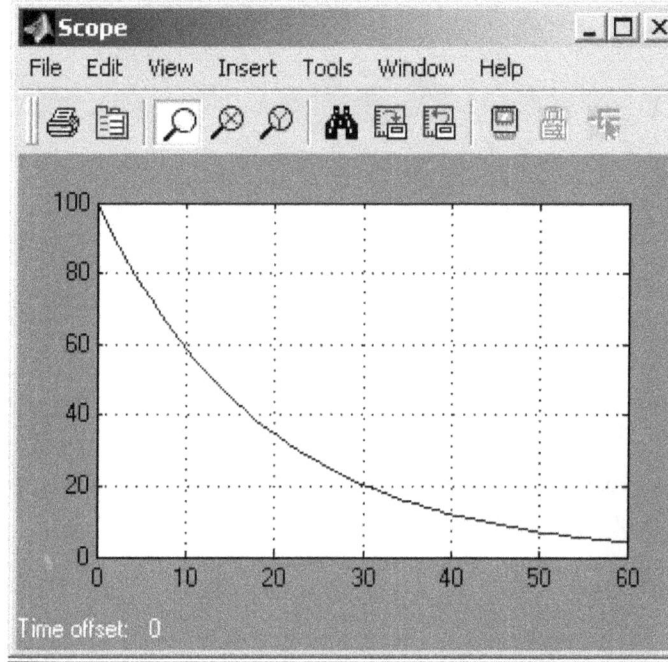

Time Constant = 6.21 sec

SIMULINK MODEL

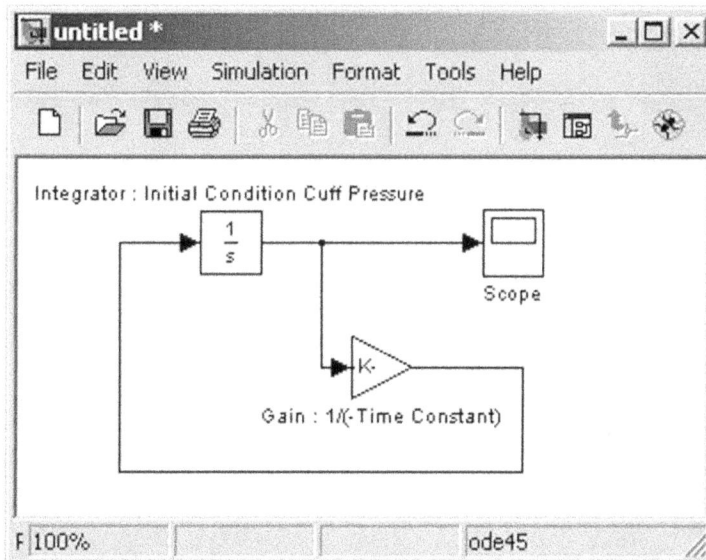

Simulnk results

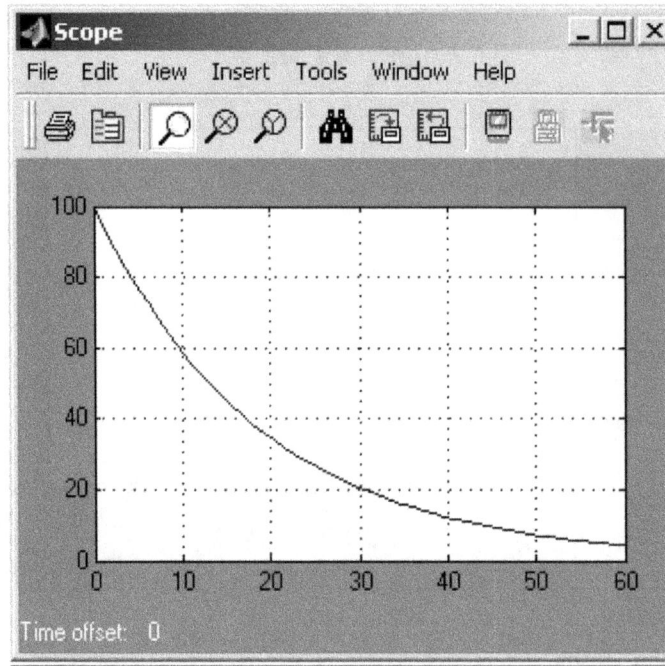

Time Constant = 18.93 sec

Fast Time Constant

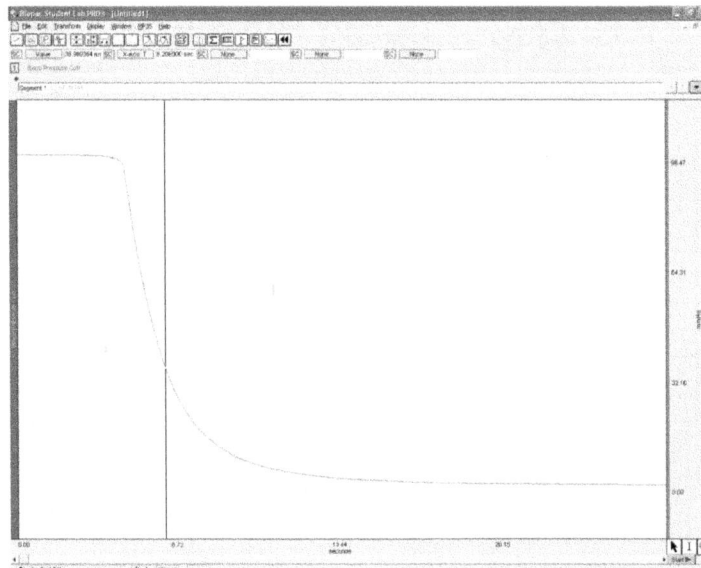

Figure Credits

LAB 9

Amplifiers

OBJECTIVE

To examine the DC amplifier from the operating-bias point of view. For this purpose, we shall also apply an AC filter to the DC amplifier so that the DC amplifier becomes easier to use in the study of AC signals.

KEY LEARNING SKILLS

- Gain
- DC bias
- AC amplification
- Linear amplification

INTRODUCTION

The basic use of amplification is so that a low-voltage signal can be increased so that it can be studied more easily. Since this course has only to deal with linear systems, our view of the amplifier is very simple in that the signal, s, is linearly scaled by amplification so that s * G = S. S shall be the amplified version of the signal s.

Ideally, amplification just multiplies a signal by a factor G = gain. In most cases, amplifiers will follow this multiplication rule and the MP36 also functions in this simple way.

Unfortunately, there are "real-world" limitations. In particular, amplifiers do not possess infinite energy to increase signals above their design limits. For example, to this end it should be easy to understand that the output voltage of an amplifier cannot exceed the voltage of the power supply.

Normally an amplifier must be "biased" such that signals will be less than the power supply voltage. If you exceed these mentioned limits, the net effect will be that G (gain) is reduced and the amplifier amplifies less than we expect.

Mathematically we define the function of an amplifier by the following linear formula:

$$Vin * G = Vo$$

Where G = gain factor; Vin = input voltage; Vo = output voltage

EQUIPMENT LIST
- BIOPAC MP36 analog to digital converter system
- Frequency generator and DC power supply
- T-connector 1
- BNC-to-BNC cable
- BIOPAC BNC-input cable (2)

BIOPAC EXPERIMENT

Amplifier Bias Point

Procedure
In this experiment you will need to measure voltage. Begin by calibrating channels 1 and 2 of the BIOPAC system for voltage. Under the set up channels menu, set the presets (BNC +50 to –50 volts) for both channel 1 and channel 2.

After calibration, connect both channel 1 (Ch1) and channel 2 (Ch2) inputs to the output of the function generator using a T-connector. Make sure that the power supply is on!

Adjust the frequency generator for the following:

A. free-run mode
B. frequency multiplier X1
C. attenuator—0 dB.
D. frequency dial = 1 Hz
E. select sine waveform

Launch the BSL *PRO* software to open a window on your PC. Open the channel menu and click the boxes on channels 1 and 2 to record and display data.

Close that window and open the BP100 acquisition menu. Set the acquisition rate to 100 Hz.

Save to **hard drive** and record **once**.

Set recording duration for 15 seconds. Verify that you can record an identical sine wave from both channels. Adjust the function generator output control so that you have a one-volt peak-to-peak voltage input to the BP30.

DC Bias Point (DC Amplifier)

Now we begin to find the DC bias point for the BIOPAC amplifier for Ch2.

Start by adjusting the sine wave on Ch2 to be max = 0.1 volt and min = 0.1 volt.

Then, leaving everything else unchanged, open the Ch2 parameter window to change Ch2 amplifier to X100 and change the Ch2 preset to (BNC +10 to –10 volts).

Start another recording and print it. You should still see two complete sine waves on Ch1 and Ch2.

If you have good sine waves, then pull out the offset control on the generator. Start a recording while you increase the offset voltage. Continue until you see that the Ch2 sine wave flattens on its top. Print this result. It is the condition showing too much +DC bias.

Do an x–y scope plot of Ch1 and Ch2 at this point and print it out.

Next, reduce the offset while you record until you just find the point where the sine wave just becomes normal again.

This is the optimal DC bias point. Print your results. Measure the offset voltage at the bias point by using the measurement cursor to find Vmax - Vmin = Vbias.

AC Amplification

Since it is generally useful to study only the AC component of a signal, we will now use this type of amplification.

First, readjust the offset so that Ch2 sine wave shows a flat top again. Then, open the Ch2 parameter window and click the AC button on. Start another recording. Print your result.

You should see the sine wave is now centered on the bias point automatically. The DC offset was "blocked" by a BIOPAC filter. Measure the cut-off frequency of the filter by adjusting the frequency to the 0.707 point using the (P-P) voltage. This is when AC (P-P) = 0.707DC (P-P).

Group Lab Report Questions

1. Submit all graphs:
    a. DC-bias point (top-flattened sine wave)
    b. Scope x-y plot for the flattened sine wave
    c. AC sine wave record
    d. Simulink of Biopac graph (a)
2. What filter does the BIOPAC system use to do AC amplification?

3. Calculate the phase shift of the BIOPAC AC filter.
4. Provide a written summary and discussion of lecture theory.
5. Since G is the slope of ch2/ch1, find the gain from your x–y plot of Ch2–Ch1 for the low- and high-voltage ranges.
6. Draw a Simulink model that models the bias.

Individual Lab Report Questions

1. It is clear from the sine wave recording that the linearity of the BIOPAC amplifier has been exceeded when the waveform no longer appears sinusoidal. That is, clipping becomes evident. Sketch how a triangle wave and square wave would appear for excessive amplification.
2. Based on your answer to question 1, what is the best way to detect nonlinear operation? The time plots or the x–y plot? Explain your answer.

Sample Lab Data

Sample Amplifiers Graph for Nonlinear Condition Input and Output Sine Waves

Input Sine wave

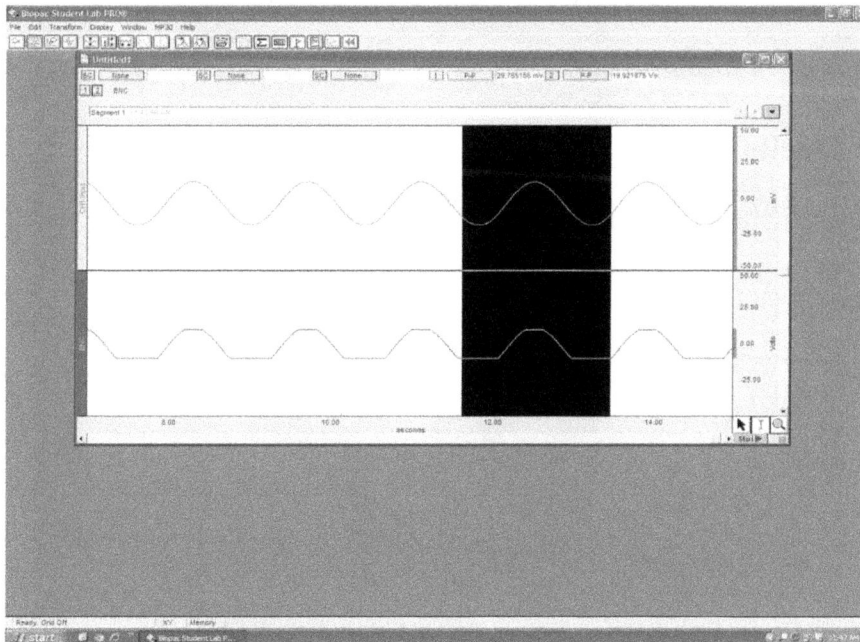

DC Transfer Function for Nonlinear Condition Vin Versus Vout
Amplified sine wave exceeding BIOPAC range

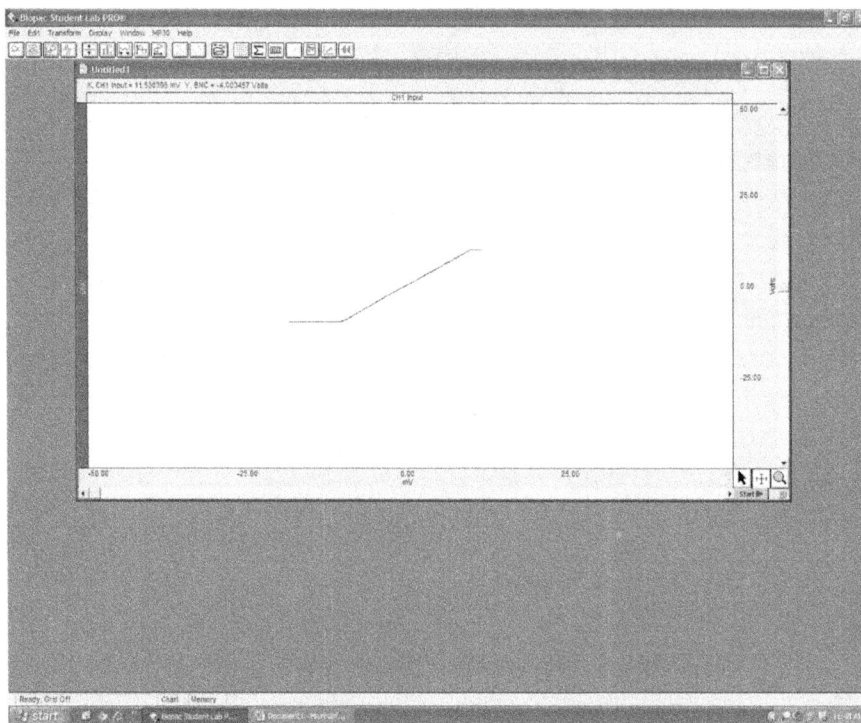

Figure Credits

Part III

Circuits

RC Frequency Response

OBJECTIVE

In this experiment we measure the sinusoidal frequency response of a RC circuit. The magnitude and phase response will be measured as a function frequency. The time response of the RC to a square pulse is also obtained to measure the time constant parameter of the circuit. The circuit is then modeled in Simulink to calculate its square pulse response. The modeled response is then compared to the actual time response data to reveal the accuracy of the circuit analysis.

KEY LEARNING SKILLS

- Bode plot
- RC filter
- Pulse response
- Cutoff frequency

INTRODUCTION

The subject of this experiment is the series RC branch circuit that is energized by a sinusoidal voltage source applied across the branch. Define the voltage output of the circuit to be the voltage across the capacitor. The circuit schematic is shown in Figure 10.1. Note that this circuit is different than the RC branch used in earlier experiments since the focus is on the capacitor voltage instead of the branch current.

Ch 1

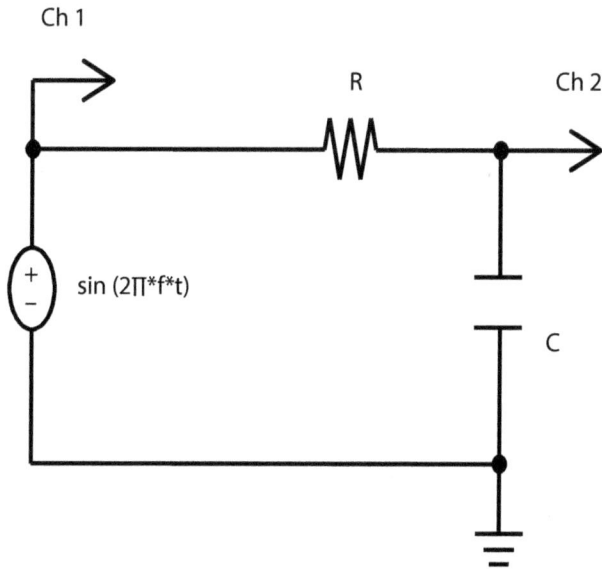

FIGURE 10.1 Electrical RC branch circuit and Set-up.

Begin the study of this circuit by first analyzing the relationship between the voltage source and the voltage across the capacitor. The capacitor voltage is simply the capacitive impedance multiplied by the current phasor, $I(J\omega)$, as,

$$V_C(j\omega) = \frac{1}{j\omega C} * I(j\omega)$$

Since there is only one loop in the circuit, we apply Kirchhoff's voltage law to find $I(j\omega)$ as follows,

$$V(j\omega) = I(j\omega)[R + 1/(j\omega C]$$

Solving for $I(j\omega)$ and then replacing $I(j\omega)$ to solve for $V_C(j\omega)$ yields

$$V_C(j\omega) = \frac{V(j\omega)}{(1 + j\omega RC)}$$

If the magnitude of the voltage magnitude of the voltage source phasor is A, then the magnitude of the capacitor voltage phasor is

$$|V_C(j\omega)| = A\frac{1}{\sqrt{1 + (\omega RC)^2}}$$

Notice that the magnitude depends on the frequency applied to the circuit. Using the real and imaginary components of the phasor we can also find the phase from

$$\phi(j\omega) = \tan^{-1}(\omega RC)$$

To simplify the analysis let

$$(\omega RC)^2 = 1$$
$$\text{also, } \omega_c = 2\pi f$$
$$f_c = \frac{1}{2}\pi RC$$
$$\text{and } |V_c| = \frac{A}{\sqrt{2}} \text{ for } \omega_c = \frac{1}{RC}$$

Also, define ω_c as the cutoff frequency, for which case, if the voltage source frequency equals the cutoff frequency the magnitude of the capacitive voltage reduced to 0.707A. Notice above that the cutoff frequency is related to the product RC. RC is also the time constant for the RC circuit. RC is the key parameter that regulates the function of this circuit. In this experiment, we will use three methods to measure RC:

Method 1. Cutoff frequency is one way to find RC from the sinusoidal response of the circuit.

Method 2. Phase is another method of determining RC from the above sinusoidal analysis by referring to the formula for phase shift. Notice that when the frequency is exactly equal to the cutoff frequency the phase shift is 45 degrees. Then, convert cutoff frequency to RC as in method 1.

Method 3. The square pulse time response. Recall that the voltage time RC response is

$$V_c(t) = V_c(0)e^{-t/\tau}$$

Where τ = RC and where $V_c(0)$ is the initial condition of the capacitive voltage the time response equation therefore provides the second means of finding the value of RC in the square pulse response.

Note: There are two experiments within this section. Both experiments measure the same RC circuit in the first trial A, the input voltage is a sine wave. In the second trial B the input is a square wave.

EQUIPMENT LIST
- BIOPAC MP36 analog to digital converter system
- Sine wave frequency generator
- RC substitution box
- BIOPAC input cable, T
- Connector clip to BNC cables and T-connectors
- Banana to banana cords as needed

BIOPAC EXPERIMENT

Rc Sinwave Frequency Response

Procedure:

1. In this experiment you will need to measure voltage. So, begin by calibrating two channels of the BIOPAC system for 0 and +5 volts.
2. After the BIOPAC system is calibrated to read voltage, wire the RC circuit of Figure 10.1. Use the RC substitution Box for R and C. Use the clip leads cables to connect to your circuit.
3. Connect one channel of the MP36 to the input voltage of the RC circuit that is the function generator output. Connect a second channel of the BIOPAC MP36 to the output of the RC circuit. In this case, take the output voltage as the voltage across the capacitor and the input voltage as the output of the signal generator.
4. Set the values on the RC box to be R = 500 KΩ and C = 1.0 μF.
5. Adjust the frequency generator for the following:
    a. Push the power button to the "on" position
    b. Push Sine wave output
    c. Frequency multiplier X1
    d. Frequency dial = 1 Hz

Launch the BSL *PRO* software to open a window on your PC. Open the channel menu and click the box on channel 1 and 2 to record and display data. Select the channel preset for a voltage BNC input.

Close the channel window and open the BIOPAC acquisition menu. Set the acquisition rate for 100 Hz.

Select save to **hard drive** and record **once**.

Type in recording duration for 30 seconds. Click the start record. Verify that you can record a 1Hz sine wave from both channels simultaneously.

Once you are able to make recordings, use the measurements menu on the Biopac window and find the peak-to-peak voltage measurement.

Adjust the function generator output control so that you have a one-volt peak-to-peak voltage input to the RC circuit. This will only be a starting voltage that is convenient to use. Should you find that your output voltage RC become too small to measure without noise, you may want to increase the input voltage. Warning: Do not change the input voltage during the course of your frequency measurements. If you do change your input, you need to redo your peak-to-peak measurements. Phase will be ok.

Now begin to collect data to measure the frequency response of the circuit by sampling the circuit's capacitive voltage peak-to-peak output at various frequencies.

Since the generator is already set to 1 Hz, measure the peak-to peak voltage on the second channel of the BIOPAC window. This is the sinusoidal magnitude of the circuit's output at 1 Hz.

Next, repeat the measurements of the circuit's input and output voltage at other frequencies. It is a good idea to start by moving the generator frequency dial to 2 Hz and then doubling again to 4 Hz, 8 Hz, 16 Hz, and so on. Record your data in a table containing frequency, input, voltage, output voltage, and phase. You will need to increase the frequency generator multiplier switch to 10x in order to reach the higher frequencies. Once you measure the output to be less than 0.1-volt stop going up in frequency. Then, start to take measurements going down in frequency. For example, 0.5 Hz, 0.25 Hz, and 0.1 Hz. Be careful to wait for a **full sine** at the low frequencies. Once you observe no voltage change, stop reducing frequency.

Next measure phase at every frequency that you just recorded, Start first at 2 Hz. You will now be measuring the relative time shift between channel 1 and channel 2. That is, between the input and output of the RC circuit.

Locate the time interval (Δt) measurement tool on the BIOPAC window. Use the measurement cursor to highlight one full sine wave on the output voltage channel. One full cycle equals the period T. Record the reading in msec. Then, using the same cursor measure the **time delay** between the input and output voltage channels. It will be a fraction of T but use the cursor to highlight the time shift between input and output. This is the time delay Δt. Notice that you see a time delay. A delay corresponds to phase lag and is therefore negative phase.

Using the following formula, you can convert from time shift to phase using your readings

$$\phi = \frac{\Delta t}{T} * 360 \text{ degrees}$$

Now repeat the phase measurement for at all of the other frequencies that you measured the peak-to-peak output and add this result to your data collection table. ADD MORE DATA TO YOUR TABLE BY ADJUSTING THE FREQUENCY GENERATOR TO FREQUENCIES BELOW 1 Hz for frequencies such as 0.5, 0.125, and lower if you still detect changes.

Be careful that you use the full sine wave to do your measurements at low frequency. Also, take care to compare the same relative points of the sine waves on channel 1 and 2. You will find that the time shift becomes very small or 0 at low frequencies according to the RC theory. At this point you should have a complete table of output peak-to-peak voltage and time shift versus frequency. Examine your table to find where the output just becomes less than 0.707 of your input voltage. The e frequency when this occurs is the cutoff frequency. You should also see that the phase shift is −45 degrees at this frequency. Using the formulas for cutoff frequency, you to calculate the value of RC. Notice that up until now you have been concerned with only sine waves and have been analyzing the circuit in the "frequency domain."

Square Wave Response

Next, we perform a measurement on the same RC circuit to observe its response in the time domain to a square wave input. Leaving everything else unchanged, adjust the frequency generator to 0.2 Hz and push in the square wave button.

Start a single recording in BSL *PRO*. You should observe a square wave on the input channel. The output channel should also have a signal that looks like a charging–discharging capacitor voltage response (sample graphs at the end of this chapter).

Next, use the cursor to measure voltage. Use it on the output wave response to find the peak-to-peak voltage. Then, move the cursor until the voltage drops to 0.368 of the initial (V peak-to-peak). 0.368 corresponds with one time constant duration.

Record the time interval for which it takes the wave to fall to .368 of the maximum value. This is the time constant (τ) for the circuit. You should now have the value of the time constant from two determinations of the cutoff frequency and the square wave method. An alert student might recognize that they already know RC from the values set into their circuit box. The circuit values of R and C then provide an additional value of the time constant. You will find that there is some difference between the measured RC and the circuit value product of RC. This difference can be explained from the fact that the frequency and square wave response is a true measurement of our total system, including the voltage source and the BIOPAC measuring system. The product RC assumes that these are the only elements in the circuit. It will turn out that this assumption is not valid and causes RC to be different from the measured RC. You will learn in later experiments that the instruments are introducing parasitic elements into your experiment. For now, it is important that you become aware that there may be "unseen" elements in your systems due to imperfect lab instruments. Later, in the Simulink model portion of this experiment you will find that your measured RC still predicts the response of your measurements very well in spite of this error.

Next, input a square wave to your RC circuit for any frequency above the cutoff frequency value that you measured. Measure the capacitor voltage response and print your data. Do the same for a triangle waveform. You can select a triangle waveform by pressing the button on your waveform generator. Can you determine what calculus operation the circuit is performing? Repeat this measurement for one frequency below the cutoff frequency. Print your results. Does the circuit accurately perform the calculus operation below cutoff frequency? Try to analyze this observation when you make your lab report.

SIMULINK

In this design experiment, the Simulink functional blocks will be used to model the square wave response of the RC circuit. Since Simulink does not provide circuit analysis features, we will use the transfer function block to model the RC circuit. The transfer function equation was derived earlier in this chapter.

Open the transfer function block in the Simulink program to see the transfer function similar to Equation 10.3. You will find that the transfer function depends on $j\omega$ in the phasor form. In the case of Simulink, you will find that the transfer function depends on S. For our purposes it will be a good approximation to assume that $S = j\omega$, in which case you can substitute the phasor transfer function that was derived earlier.

Hel result. Then, compare your Simulink model graph to your BIOPAC recorded square wave response. If necessary, adjust the square wave model parameters until you are able to model your data as best as possible. Repeat the model calculation for a frequency above the cutoff frequency. Make sure that you find that the model is then performing the integral of the square wave.

Group Lab Report Questions

1. Design an alternative Simulink model that directly solves the loop equation for the RC circuit without using the transfer function.
2. Submit graphs for all trials performed.
3. Provide a bode plot using your frequency-magnitude and phase data. Sketch the asymptotes of the transfer function on the same plot.
4. Use the time constant obtained from the square wave experiment to calculate the cutoff frequency.
5. Find the cutoff frequency from your bode magnitude and phase plot. Compare values with question 4 and your measured cutoff frequency. Do an error analysis between each value obtained.
6. Prepare a summary of your experiment. Discuss how your lecture course applies.
7. What kind of filter is this RC circuit? Bandwidth =?
8. Design an RC filter circuit that reduces the amplitude of a 3 Hz sinusoidal input wave to 0.25 of the input amplitude.
9. What calculus operation did you find that this circuit provides? Prove it using circuit phasor analysis. Why did it only work for frequency above the cutoff?

10. Sketch the bode plot for the RC circuit transfer function. Draw the asymptotes on the bode data plot in question 3.

Individual Lab Report Questions

1. Use the transfer function of the RC circuit of this lab to draw the bode magnitude plot.
    a. Does the break frequency correspond with your measured plot?
    b. Does your measured attenuation rate correspond with –20 dB/decade?
2. At high frequencies, you should have observed that the square wave is integrated by the RC filter.
    a. Why does the filter not provide a triangle wave at low frequencies?

Sample Data

Lab 9 – RC Frequency Response

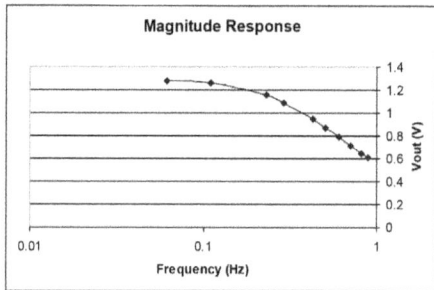

Vin	Vout	Freq (Hz)	td(S)	Phase (deg)
1.92	1.278	0.062	0.575	-12.834
1.92	1.26	0.11	0.39	-15.444
1.92	1.16	0.23	0.273	-22.6044
1.92	1.09	0.288	0.429	-44.47872
1.92	0.95	0.428	0.327	-50.38416
1.92	0.872	0.507	0.291	-53.11332
1.92	0.788	0.603	0.279	-60.56532
1.92	0.714	0.7074	0.243	-61.883352
1.92	0.641	0.815	0.194	-56.9196
1.92	0.606	0.894	0.174	-56.00016

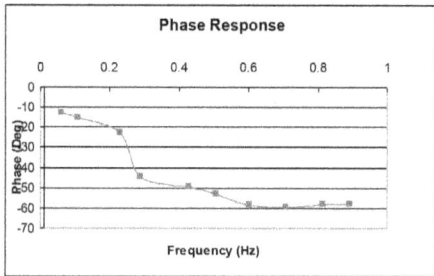

Square wave response

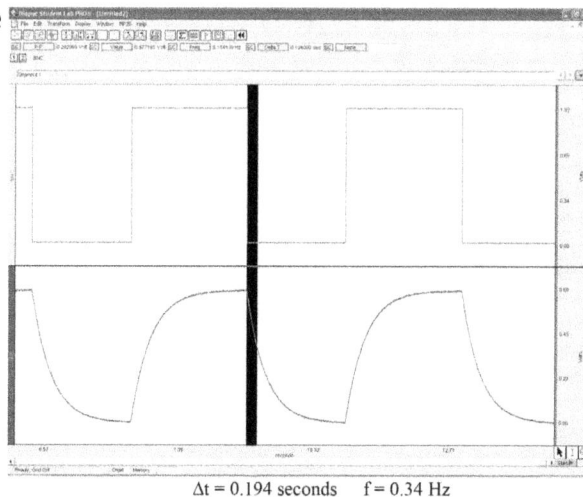

Δt = 0.194 seconds   f = 0.34 Hz

Square wave input

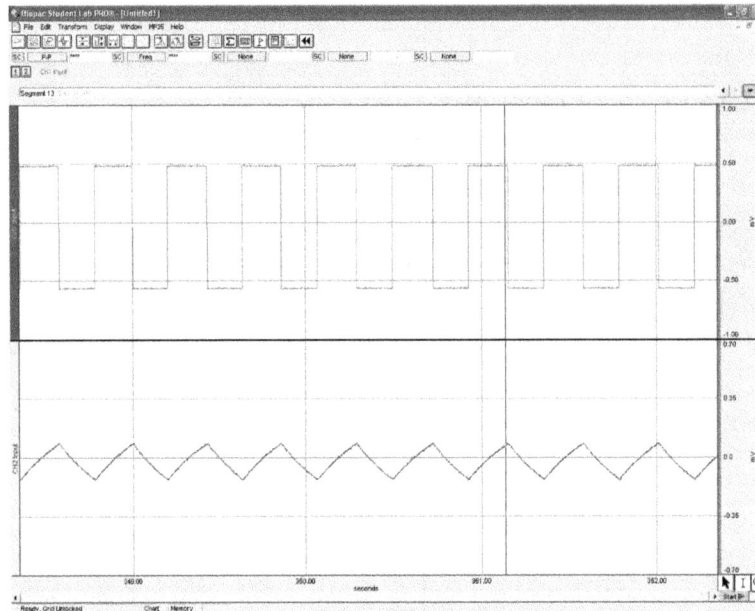

Take at Frequency = 2 Hz

RC square wave

Simulink RC transfer function

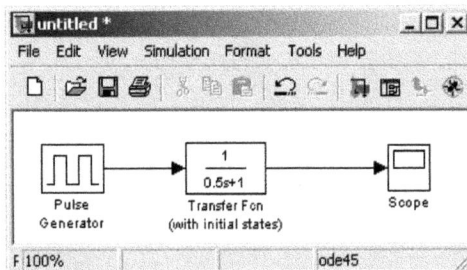

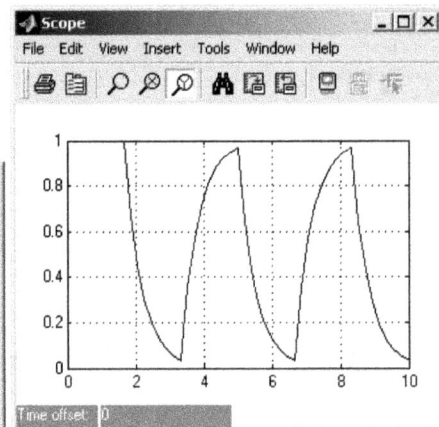

Figure Credits

LAB 11

Time-Dependent Electrical Variables and Sources—Phasors

OBJECTIVE

In this experiment, we will apply the concept of phasor analysis to the sinusoidal testing of some basic electrical elements.

KEY LEARNING SKILLS

- Phasors
- V-I Plot
- Electrical Elements
- Phase Plot

INTRODUCTION

The Phase (i-v) Plot

In this experiment, we will use an analysis method that allows us to examine system variables while time is treated as an implicit variable. The value of this approach is that it is a simplifying technique in that we don't need to be directly concerned with the effect of time on any variable.

Consider the example below where we measure the voltage and current in a single resister. The experiment circuit is shown here in Figure 11.1. The electrical element in this case is a resistor of 1KX, but any electrical element could have been tested in place of the resistor. In this experiment, we first examine a resistor and then we replace it with a capacitor.

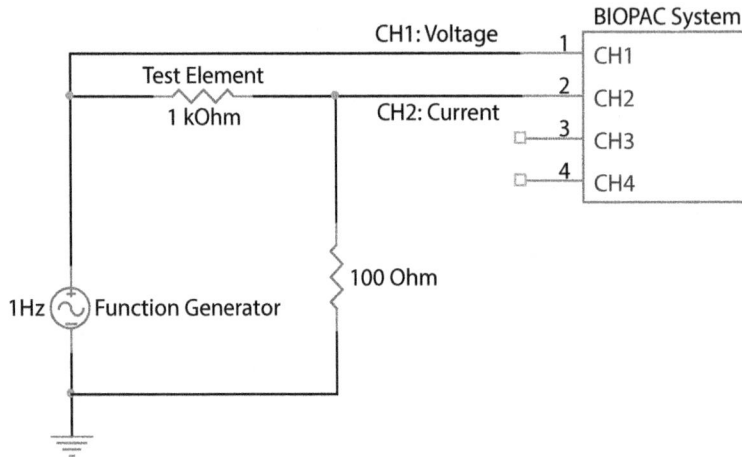

FIGURE 11.1 Resistor and Source Circuit.

Define the voltage source to be v(t) = sin (ωt)

Find the series current as: I(t) = Ch2(t)/100.

Notice that both current and voltage are functions of time in this circuit. If we plot the voltage and current separately, we see that they both follow the sinusoidal source voltage. See Figure 11.2.

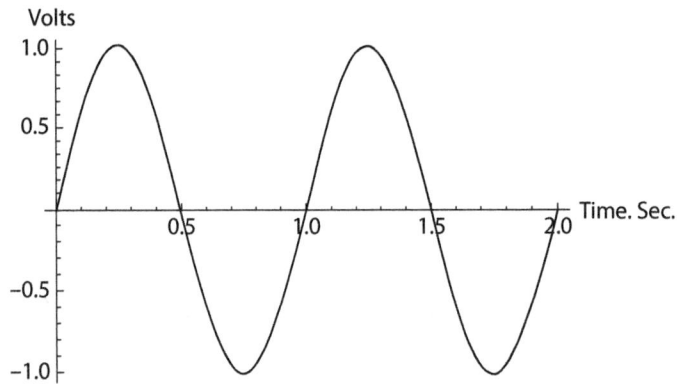

FIGURE 11.2 Sinusoidal voltage source Sin(2πt).

For example, in Figure 11.3, current I(t) is shown to be a sinusoid.

Current mA

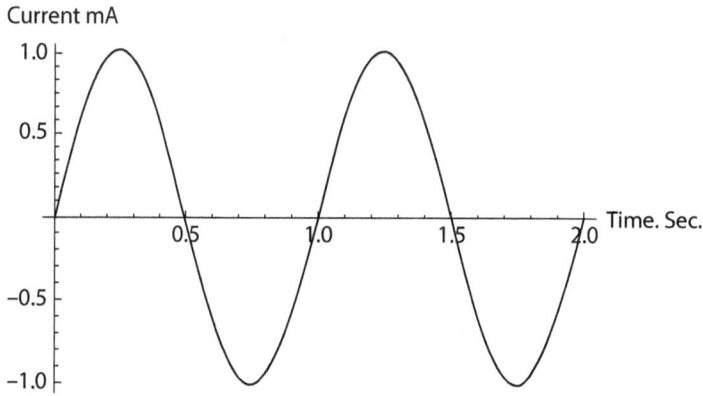

FIGURE 11.3 Loop current in Figure 1 for R = 1 kOhm and voltage of Figure 2.

Now examine the phase plot of voltage versus current below in Figure 11.4.

Note: This graph plots the instantaneous magnitude of voltage against current. Note also that the angle between the current and voltage vector is the phase difference and is 0.

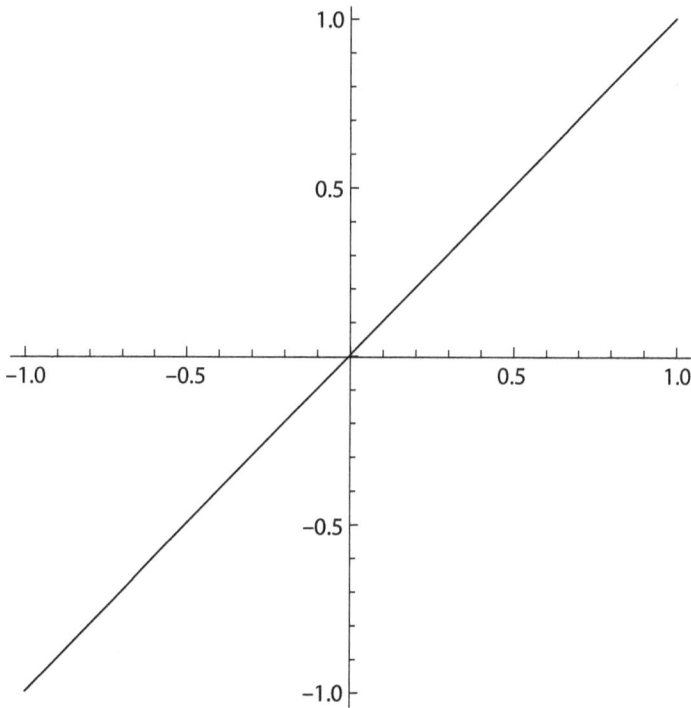

FIGURE 11.4 i-v phase plot for a resistive element.

Notice that the result is very simply a straight line due to the 0 phase difference. The slope of this line is equal to the value of R. Hence, we find the simplifying feature of the phase plot.

It should be clear that since each point on this plot is implicit in time that the current and voltage sinusoids are exactly in phase since they are directly proportionate to each other according to Ohm's Law.

Capacitive Test Element

Now, repeat the same analysis using the circuit below in Figure 11.5.

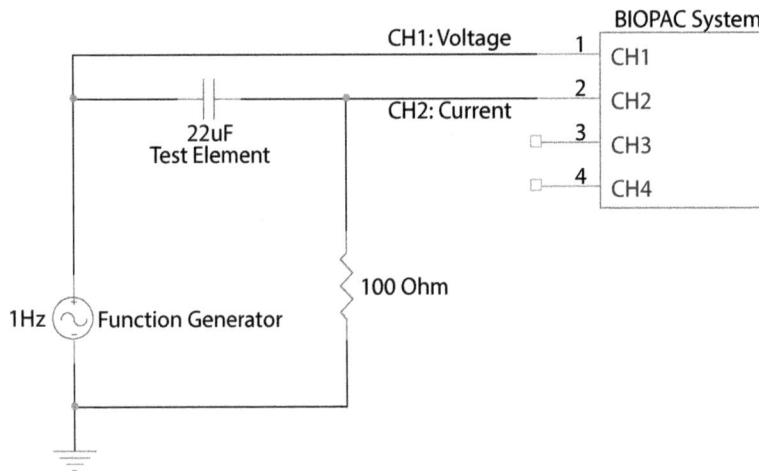

FIGURE 11.5 Capacitor element circuit. Use frequency = 70 Hz.

Note that the circuit is similar to Figure 11.1; the test element, 1kOhm resistor, has been substituted with a 22uF capacitor.

The current voltage law for capacitance C is

$$\frac{dv}{dt} = \frac{i}{C}$$

And in this example, use instead C = 2F.

Again, if the current measuring resistor is small, it will not disturb the circuit and almost all of the voltage will be across the capacitor.

We can solve for i by substituting the voltage into the capacitor i-v law and find that the current is

$$i = 2\text{Cos}(\omega t)$$

The current and voltage of the capacitor is plotted below in Figures 11.6 and 11.7 with t in radians:

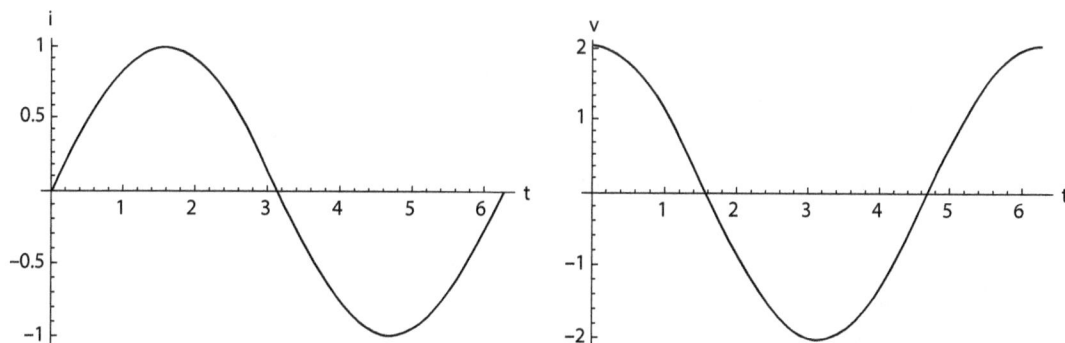

FIGURE 11.6 and 11.7 Current and voltage of the capacitor circuit.

Notice that the phase plot, in Figure 11.8, is now ellipsoidal as opposed to a straight line. This indicates that i and v no longer follow each other directly in time but instead are delayed. Referring to the separate time graphs of each. In this case we see that the voltage leads the current in time. Each point on the i-v ellipse graph provides an angle with the origin, showing the instantaneous phase.

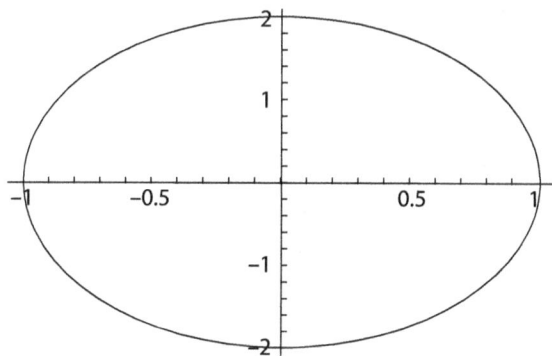

FIGURE 11.8 i-v phase plot for a capacitive element.

EQUIPMENT LIST

- Matlab Simulink
- BSL *PRO* software
- BIOPAC system
- Function generator
- BIOPAC BNC adapter cable
- RC substitution box

- 1 BNC-to-BIOPAC cable
- 2 BNC T-connectors
- 2 BNC to clip adapter cables

BIOPAC EXPERIMENT

1. Wire the set-up as shown in Figure 11.1. Then continue to step 2, 3, and so on. After the response for the resistive element is complete, remove the 1kX resistor and configure the circuit to what is shown in Figure 11.5; once again complete the prescribed steps to result in the above phase plots. For both set-ups use a 1V P-P generator, at 70 Hz, as the input voltage.
2. Plug in the BNC cable to the channel 1 of MP36 and the function generator. The channel 2 BNC cable is connected to the clip cable that is placed across the 100W resistor.
3. Open BIOPAC BSL *PRO*.
4. Open the menu MP36 and use these settings (#5 and #6 below):
5. Go to set-up channel and input the following data:
    a. Preset: BNC (–10V to +10V)
    b. View/change parameters: USE DEFAULT SETTINGS
6. Setup acquisition:

    Record and save once using the hard disk.

Sample rate	1000 samples/second
Acquisition length	30 seconds

7. On the function generator, push in the sine wave button on the top right-hand side.
8. Adjust the frequency such that it reads 70 Hz.
9. When you press the start button in BSL *PRO* you should receive a graph that looks like a sine wave. If not, press the resize button so that it resizes the graph to fit the screen.
10. Use BSL *PRO* to measure the output voltage into channel 1 (Ch1) and channel 2 (Ch2). Set your acquisition and gain parameters to obtain pure sine waves on both channels. If necessary, you may increase the generator voltage as needed to provide good signal to noise. Note, that in prior labs you have only used a single Ch1 to measure data. In this experiment you will use two channels simultaneously. At this point, be sure to go to the channel set-up menu and select Ch2 to record and plot data. Use the channel gain as the same as channel 1. Both channels will use the BNC 10-volt inputs.
11. In order to convert Ch2 into current, you must divide by 100. Go to: *transform, waveform math*, **Ch2/k = Ch2**. The variable **k = 100** for the current in the 1000 X resistor. Here **k = 100** for the both the capacitor and resistor circuits. Note that since the current resistor is small, we assume most of the voltage is across the test element.

12. To complete an x–y analysis procedure (*i-v* plot), click the third icon on the top left. Be sure that you have selected current versus Ch1 new to be displayed. Preferably you would put current on the x-axis.

13. Calculate the slope of the resistive element. This is done by dividing the (P-P) values of Ch1 and Ch2 from the original sinusoid plot. The slope should be 1000 for the resistive element linear phase plot. The slope on the capacitive *i-v* plot cannot be measured since it is changing with time.

14. Change your circuit that of Figure 11.5 (capacitor element circuit figure) and repeat the above measurements for the capacitive circuit. Print your voltage and current graphs. Keep the generator at 70 Hz.

15. Repeat the *i-v* plot measurement for the capacitor again for a sine- and a square wave. For the sine wave, measure the magnitude and phase for *i-v* waveforms. Print your graphs. Note that the *i-v* plot remains a loop for each wave. Also, notice that the current wave is the derivative of the voltage. You do not need to measure phase in the case of the square wave. It is important to realize that you have just designed a derivative computing circuit as part of your experiment.

SIMULINK

Design a Simulink model that represents the BIOPAC experiment. Reproduce the phase plots for both the capacitive and resistive electrical elements.

Examine the outputs for the two plots and be sure to reproduce the experimental results from the BIOPAC experiment as best you can. Pay attention to frequency and sample rate. Hint: Use the x–y scope from the sink library to get an x–y phase plot in Simulink. Then generate the $v = x$ and $i = y$ signals separately from a sine source. Note: The x–y scope does not auto scale. So, you will need to manually alter the x–y axis range to fit your data. Just be concerned with the phase plot shape.

Group Lab Report Questions

1. Provide a copy of all graphs from both the BIOPAC experiment and Simulink results.
2. Calculate the phase shift between voltage and current observed for the capacitive element.
3. Find the magnitude of the capacitive voltage and current from your data. Using the magnitude (P-P) and phase data, write the voltage and currents in phasor notation.
4. Calculate the phase angle from the magnitude data. Does it agree with your measured phase angle? Why or why not? Explain in detail with references from your lecture text.

5. Prepare a summary of your experiment. Please note that this summary **MUST** encompass the theoretical and experimental aspects that were expressed in this experiment. Equations and background from the lecture text must be properly referenced.
6. How does the capacitor current change with frequency? Derive a relationship.

Individual Lab Report Questions

1. What would the current waveform look like in the capacitor experiment if you used a square wave instead of a sine wave. Provide a sketch.
2. What would the current waveform look like in this experiment if you used a triangle wave instead of a sine wave. Provide a sketch.
3. If you used an inductor L in place of the capacitor in this experiment derive an equation for the current.

Sample Lab Data

Resistor

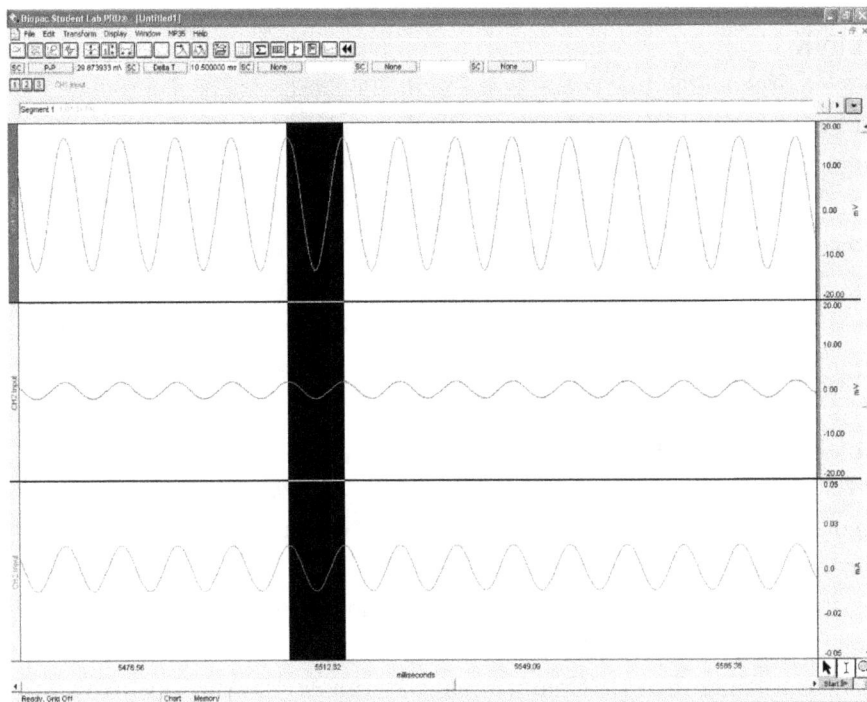

100 Hz Sine wave Source Voltage and Measured Current Across 150Ω Resistor

$$V_S = 30 \sin(200\pi) \text{ mV} \qquad V_R = 3.88 \sin(200\pi) \text{ mV} \qquad I_R = 0.026 \sin(200\pi) \text{ m}$$

i-v Plot for Resistor

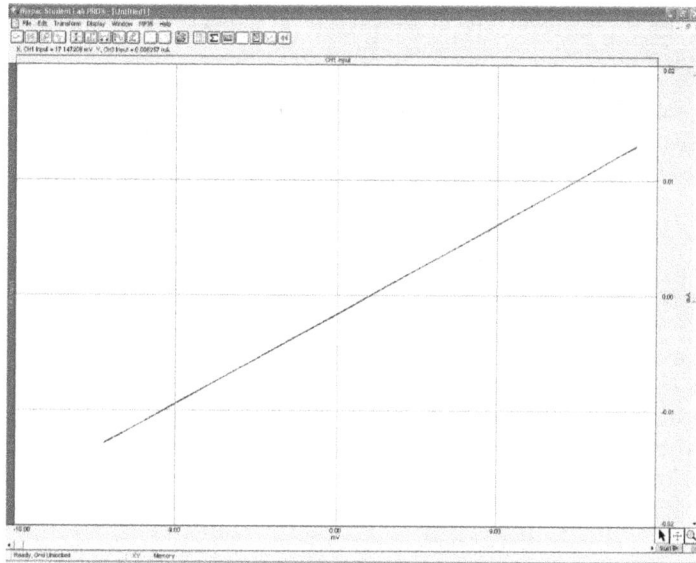

Capacitor

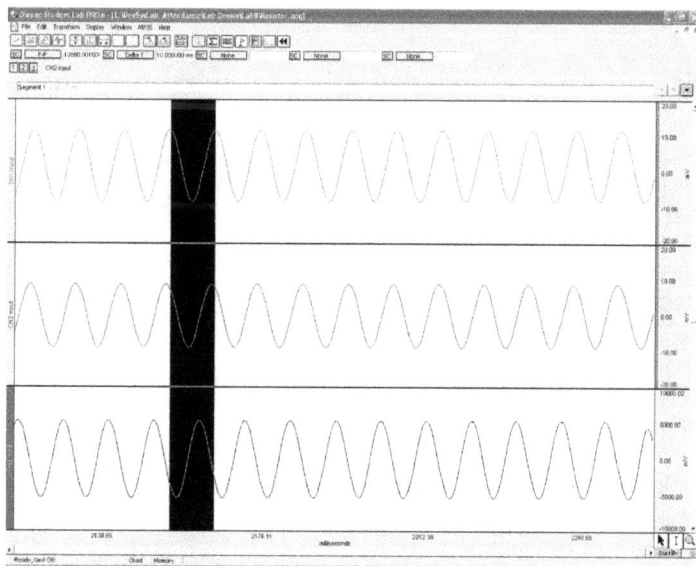

100 Hz Sine wave Source Voltage and Measured Current Across 22uF Capacitor

$$v_s = 20 \sin(200\pi) \text{ mV} \qquad v_c = 17.65 \sin(200\pi) \text{ mV} \qquad i_R = 10860 \cos(200\pi) \text{ mA}$$

$$\Delta t = 6.5 \text{ mS}$$

i-v Plot for Capacitor

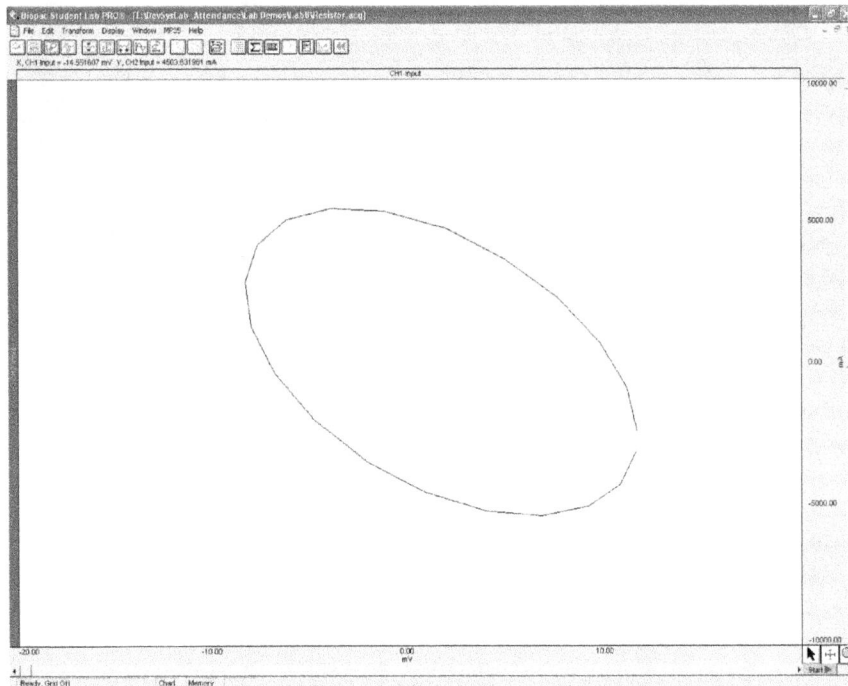

SIMULINK

Resistor

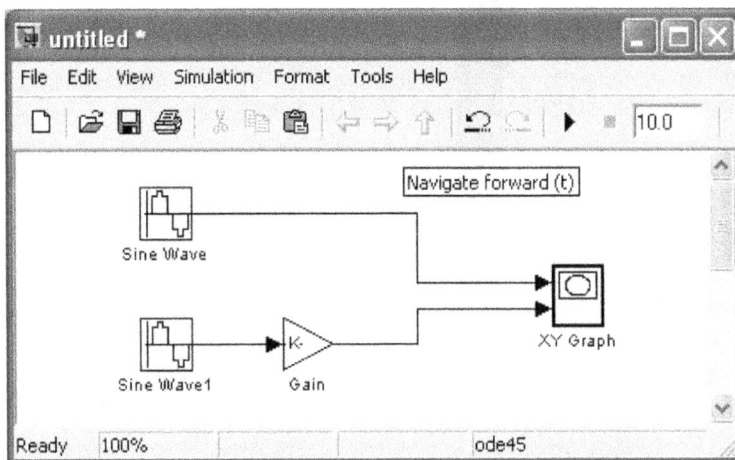

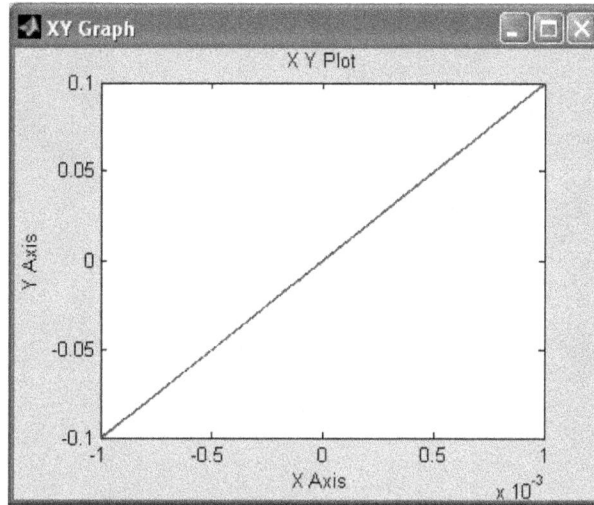

Capacitor

Simulink model and I-v output.

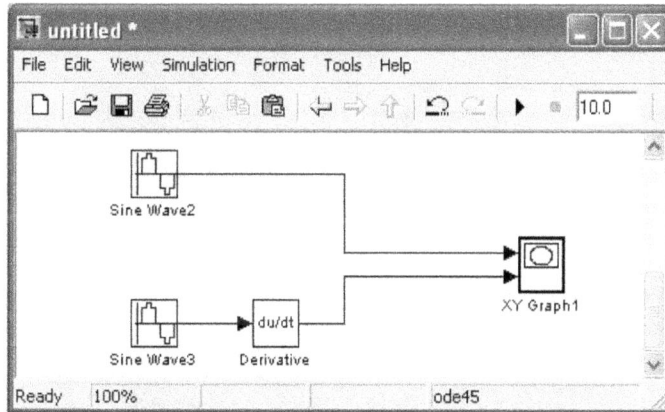

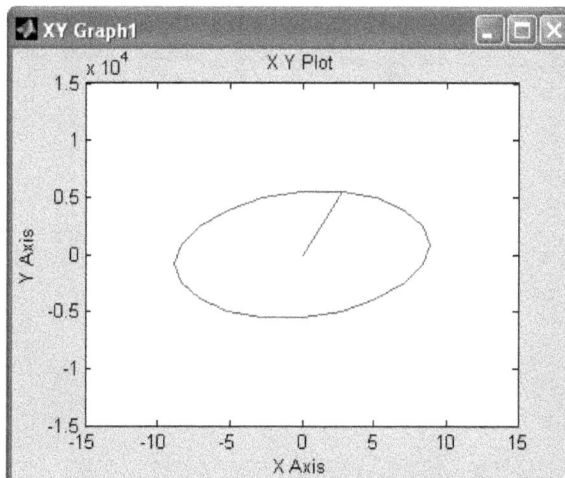

Figure Credits

LAB 12

Thevenin Equivalent Circuit

OBJECTIVE

To study the use of the Thevenin equivalent circuit and its applications to gain a better understanding of our lab's function generator.

KEY LEARNING SKILLS

- Thevenin source circuit
- Open circuit
- Short circuit
- Thevenin impedance
- Maximum power transfer

INTRODUCTION

The general circuit for the Thevenin equivalent circuit model is provided below in Figure 12.1.

The idea behind any equivalent circuit is that a simpler well-understood circuit (the equivalent circuit) can be used to replace a more complex circuit between two nodes of the circuit.

In applying this circuit simplifying technique, we are requiring only that the voltage and current external to the equivalent circuit are identical at the two nodes.

Given this definition, it should be clear that the equivalent circuit may be a completely different circuit between the nodes that is being replaced by the Thevenin circuit. In this lab, we will be using the Thevenin circuit to model the waveform generator output circuit.

As you might expect, the function generator internal circuitry is much more complex than that of Figure 12.1.

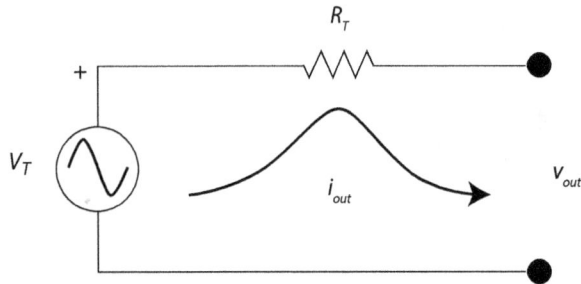

FIGURE 12.1 General Thevenin circuit model

There are only two parameters that must be known in order to describe the Thevenin equivalent: the Thevenin voltage and the Thevenin resistance (or impedance).

A simple two-step procedure can be used to discover these two parameter values from external circuit variables.

First, we measure the no-load or open-circuit voltage.

As you see from Figure 12.1, if there is no external current, then the internal Thevenin current must also be 0. In this case, the voltage at the external nodes must be equal to the Thevenin voltage V_{th}, since there is no voltage drop in the Thevenin impedance. Now that V_{th} is known, the Thevenin impedance Z_{th} can be found by applying an external load.

In the simplest case we can apply a short circuit to the Thevenin circuit. This yields the short circuit current I_{ss}. Then by Ohm's Law, we can solve for Z_{th} as follows:

$$Z_{th} = \frac{V_{th}}{I_{ss}}$$

Now, with Vth and Z_{TH} known, the function generator can be replaced with its Thevenin equivalent circuit in all future experiments.

EQUIPMENT LIST
- PC with BIOPAC system and software
- Simulink software
- RC substitution box
- Function generator
- 1 BIOPAC BNC input cable, T-connector, and clip connector

BIOPAC EXPERIMENT

Procedure

Begin by calibrating the MP36 to read the voltage on channel 1.

Then, connect one channel of the MP36 to the circuit in Figure 12.2.

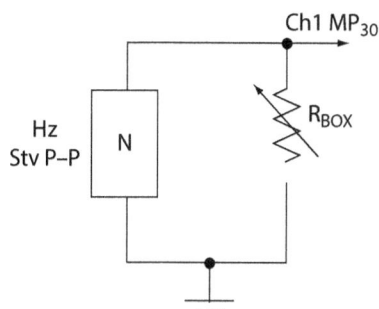

FIGURE 12.2 Lab circuit.

Set the BSL *PRO* software to the following settings:

- Channel gain X100, DC input mode
- Acquisition rate of 100 Hz

Save to hard drive once. Record for 15 second duration.

Set channel 1 to acquire data and plot; set R_{BOX} = 10 MΩ.

Adjust the function generator to 1.0 Hz and adjust the output so that you can record a 5-volt peak-to-peak sinusoid.

(The data that you record in this experiment will help you to calculate Z_{th} and V_{th}.)

Create a table of voltage measurements by varying the R_{BOX} down from a 100 Ω starting value to 10 Ω by decrements of 10 Ω.

For each new value of R, use the BSL *PRO* software to measure the peak-to-peak voltage and add this value to your table of R and V.

Lastly, remove resistance from the circuit entirely. Note that now the MP36 input is connected directly to the waveform generator. Observe how the absence of resistance affects the circuit.

Measure the peak-to-peak voltage at 10 and 100 Hz. For this condition, notice that the only load on the waveform generator is the MP36.

During most of our experiments, we assume that the MP36 input current is minimal. In fact, it is near only 1–10 micro amps. For one volt this is the same as a 1 MΩ load.

Hence, let's assume that this is equivalent to open-circuit conditions, providing us with V-Thevenin.

With R (box) removed, measure the peak-to-peak voltage at 10 and 100 Hz and record the values. Notice that we have not measured the short circuit condition yet. But you will recall that we measured the output of the generator with a 10 W load. In this case we will assume that the 10 Ω load is much smaller than the Z_{th}. Therefore, we assume that it approximates a short circuit condition. Solve for $R_{thevenin}$, using Ohm's Law.

SIMULINK

For the Simulink Model it is required to create a model the predicts the MP36 input voltage given the $V_{thevenin}$, $R_{thevenin}$, and R_{box}, referring to Figure 12.2 (above). You will find that the MP36 voltage is really the output of a voltage divider circuit.

So, using the voltage divider rule you can find that

$$V_{MP100} = \frac{V_{thevenin} : R_{box}}{R_{thevenin} + R_{box}}$$

(For further reference, refer to your classroom text.)

You will need to input a sinusoidal Thevenin voltage source of the same magnitude and frequency as in the BIOPAC experiment.

Use the Simulink blocks to calculate the MP36 input voltage using the above voltage divider formula.

A working model should give you a sinusoid of magnitude that follows your table of measured R and V according to the values in the divider equation.

Use the $V_{thevenin}$ and $R_{thevenin}$ that you measured in the BIOPAC experiment.

Then vary the value of R_{box} to predict the same measurements that you obtained for R_{box} using your Simulink model. Record the values of the MP36 voltage for the five values of R_{box} that you did in the BIOPAC experiment. Just measure the peak-to-peak values of V_{MP36}.

Group Lab Report Questions

1. Print a graph for a single value of R(box) for the BIOPAC system and Simulink. Create a graph of voltage (peak-to-peak) versus the box resistance for both frequencies measured.
2. Plot the MP36 voltage that was obtained from the Simulink experiment on the graph from question 1. Do an error analysis and discuss sources of discrepancy. Does frequency affect the result? Is Z Thevenin impedance or just resistance? Explain.
3. What was $V_{thevenin}$ and indicate it on the resistance graph. Assuming that the 10 Ω resistance is a short circuit condition, use this condition to find $R_{thevenin}$ from the theory and using Ohm's Law.

4. Given the values of V and R Thevenin and the circuit model of Figure 12.1, perform a loop analysis to solve for the output voltage input to the MP36 for any value of load resistance R(box). Plot this equation and compare your result to the data in answer 1.
5. Provide an error analysis for question 4.
6. What R_{box} value caused the output voltage to be 0.5 $V_{Thevenin}$? Show this from a loop analysis of the Thevenin circuit.
7. Describe another method that uses question 7 to find the Thevenin resistance.
8. Derive a formula that predicts the MP36 voltage as a function of R Thevenin and R(box).
9. Provide a written summary that discusses what you learned in this experiment relative to your lecture course material. Did the experiment validate the theory?
10. If the input resistance of the MP36 is 1 MΩ, show how this will affect your results. Provide a new circuit model using this information.

Individual Lab Report Questions

1. Use your data to show that the maximum power in the resistor box occurs at or near the Thevenin resistance.
2. Since the function generator Thevenin resistance was about 50 ohms, was it good to assume that this value need not be considered in our previous experiments? Provide some supporting calculations for your answer.

Sample Lab Data

THEVENIN EQUIVALENCE

Frequency = 100 Hz		
RBOX (Ω)	Vpp(simulink)	Vpp(BIOPAC)
100	3.31	3.32
80	3.05	3.12
60	2.7	2.92
40	2.2	2.15
20	1.4	1.37
10	0.82	0.98

OPEN CIRCUIT VOLTAGE

Frequency	Vpp
10 Hz	5
100 Hz	5

SHORT CIRCUIT CURRENT

V(10Ω)	Iss
0.98	0.098
0.98	0.098

THEVENIN EQUIVALENCE

Zth
51.02041
51.02041

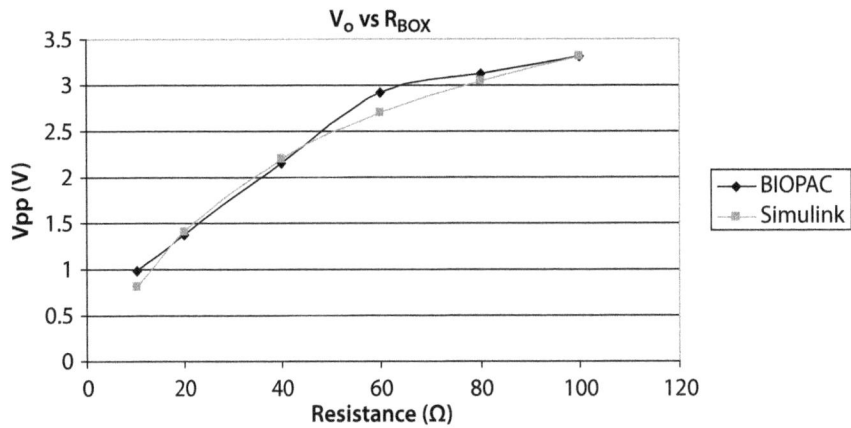

V_o vs R_{BOX}

Simulink

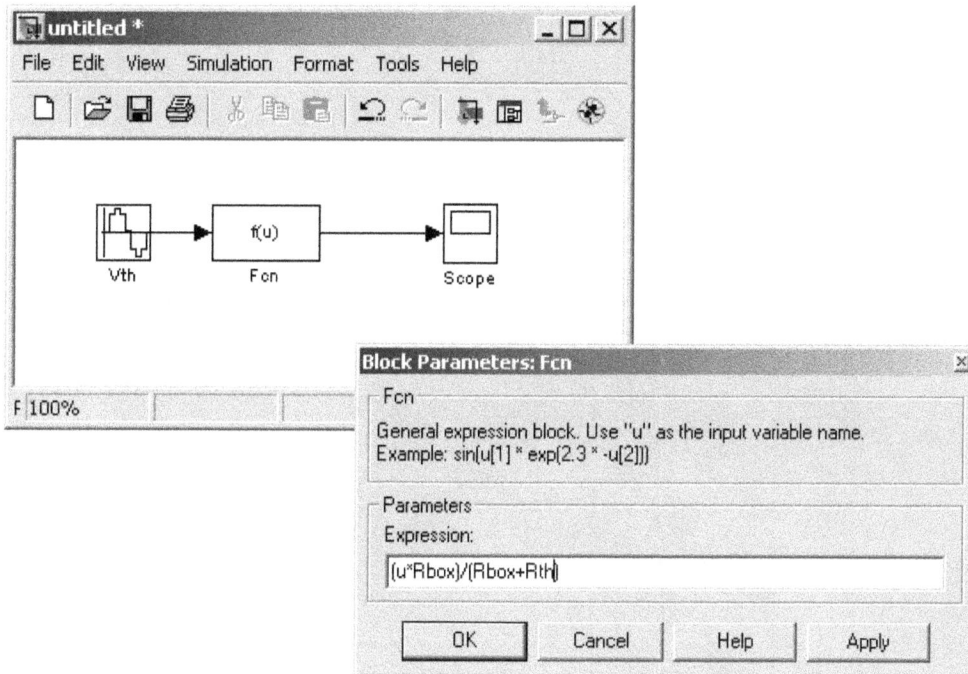

Figure Credits